Praise for Truth from the Valley

"The post-digital era is coming. Are you prepared to lead your organization there? Find out. *Truth from the Valley* gives readers a window into the critical competencies of high-performing organizations of the next decade."
Trevor Schulze, CIO, RingCentral

"As Winston Churchill once said, 'Tact is the ability to tell someone to go to hell in such a way that they look forward to the trip.' While not explicitly advocating such extreme measures, *Truth from the Valley* tactfully provides its readers with the survival skills they need to navigate the ever-changing IT landscape. Here you can tap into the wisdom of a successful serial CIO that has not only survived but thrived where others have struggled."
Declan Morris, Former CIO, Splunk

"IT leaders can put away their Taro cards and Ouija boards. There's no need to guess what the future holds. Just read *Truth from the Valley* for a peek into the critical competencies that IT teams need to win in the future. Mark Settle helps his readers work their way through initial reactions of denial, anger, bargaining and depression to arrive at a point where they're prepared to take the actions needed to empower their teams, themselves and their companies. A must-read for the leaders of tomorrow."
Yvonne Wassenaar, CEO, Puppet

"Canary environments are test environments in which changes to software systems are initially tested on a limited number of users. Silicon Valley is a canary environment for the next generation of IT management practices in the modern enterprise. *Truth from the Valley* leads its readers on a guided tour of this canary environment and leaves it to the readers to determine which practices are applicable to their companies and their teams. An indispensable planning guide for any leader who has been asked to present a strategic roadmap for the IT function during her next annual budget presentation."
Alexander Rosen, Managing Director and
Co-Founder, Ridge Ventures

"Mark Settle is one of the few people who really understands how IT is evolving in the enterprise – he's been at the front-lines as a seven-time CIO for the past 25 years. Mark's experience as a CIO has run the gamut from running IT at one of the largest oil and gas companies to an iconic financial services company to one of the fastest-growing technology players. In *Truth from the Valley*, Mark shares fundamental truths regarding the irreversible changes occurring within the IT industry. This book provides practical advice on how to adapt to these changes and position IT teams for success in the coming decade."

Sid Trivedi, Partner, Foundation Capital

"In Silicon Valley, we closely guard secrets about how to build high growth tech companies. Mark has done a service for everyone living outside the Valley by revealing what's unique, what's hard, and what's misunderstood about supporting the technology that supports the technologists. I have a shelf full of Silicon Valley anthropologies but none explains how we actually operate like *Truth from the Valley*. This should be required reading for every employee at a venture-backed startup – not just CIOs!"

Dan Turchin, CEO, Astound

"I thoroughly enjoyed reading *Truth from the Valley* – it totally captures the opportunities and challenges every CIO is facing today. Mark Settle is part IT Yoda and part CIO Therapist. His advice is practical, honest and actionable. IT leaders who fail to heed his advice do so at their own peril!"

Julie Cullivan, CIO, Forescout

"Mark Settle's wealth of experience as an accomplished CIO in Silicon Valley provides him with the ideal perspective to opine on the tumultuous future facing Information Technology executives everywhere. IT careers in the 2020s will not be for the faint of heart, but Mark graciously shares his practitioner's wisdom on emerging tech challenges and opportunities."

Bill Miller, CIO, Netapp

"*Truth from the Valley* is a must-read for IT leaders around the world who are interested in their personal growth and strategies for keeping their teams relevant."

Tony Young, CIO, Sophos

"Too many IT leaders are locked into traditional ways of managing people, processes and technology. They need to read *Truth from the Valley* to escape the insanity of doing the same things over and over again and expecting different results. This book is a much needed dose of therapy for IT teams that are tired of being middle-of-the-road performers and truly seek to become leading edge organizations."

Chris Borkenhagen, CIO, Docker

"*Truth from the Valley* provides a template for IT leadership in the next decade that has been forged within the startup community of Silicon Valley. The revolution in management practices that has occurred within the Valley is one of the least researched and yet most impactful phenomenon altering the ways in which modern enterprises exploit information technology. This concise book defines the next generation IT leadership model based upon clear, actionable strategies for adopting people, technology and organizational processes that have been tested and proven in Silicon Valley. If every reader simply embraces One Idea presented here, the better their enterprises will be!"

Stuart Evans, Distinguished Service Professor,
Carnegie Mellon University – Silicon Valley

"If there's a key takeaway from our work at the intersection of 'suits and hoodies' it's that the IT industry has experienced fundamental changes during the past five years. These changes are radically altering the ways in which IT will deliver business value in the future. Practitioners can either lead, follow or get out of the way of the current revolution. If you want to lead, read *Truth from the Valley*!"

Jonathan Lehr, Co-Founder & General Partner of Work-Bench and former Morgan Stanley Office of the CIO

"*Truth from the Valley* zeroes in on a set of complex, novel challenges facing IT leaders today. This book is a practical playbook for success in an increasingly dense and disrupted tech ecosystem with actionable advice for both IT leaders and the technology vendors that seek to engage them."

Shruti Tournatory, VP, Portfolio Growth, Sapphire Ventures

"IT is entering a new age and organizations that succeeded in the past can't rest on their laurels and assume success during the 2020s. The ones that will survive are the ones that are willing and able to adapt. Mark Settle's futuristic views in *Truth from the Valley* are evidence that he just might be the Charles Darwin of the IT world."

Casey Renner, Executive Network Director, OpenView Venture Partners

"Mark Settle is perhaps the best person to write about the trials facing Silicon Valley tech leaders due to the unprecedented pace of innovation occurring there. As a tech chief at Fortune 500 companies multiple times over and as the former CIO of a Silicon Valley security company, he's eminently qualified to predict the downstream impact of Silicon Valley management practices on companies everywhere. Read this book as a working manual on what is coming next!"

Peter High, President, Metis Strategy and Author of *Implementing World Class IT Strategy*

"Mark Settle delivers yet again with his incredible wit and invaluable wisdom for technology leaders. *Truth from the Valley* highlights the organizational capabilities that IT teams need to survive and prosper in the 2020s. Every IT leader needs to join Mark's One Idea Club. I'm already a member!"

Alvina Antar, CIO, Zuora

"I have always been a big believer in the wisdom of the technology community and its willingness to share learnings and perspectives with one another. Mark Settle is the consummate community collaborator. His willingness to share insights and seek advice is a wonderful strength that should be emulated by every IT leader. I personally benefit from spending time with Mark and always come away from our conversations having learned something new. You can benefit as well by reading *Truth from the Valley*."

Paul Chapman, CIO, Box

"Today's rapidly shifting technical landscape demands continual re-education, constant innovation, and persistent reimagination of what's possible. *Truth from the Valley* is a richly detailed book that looks deep into the changes that are currently disrupting IT organizations. Mark Settle offers practical and actionable insights into the ways in which next generation IT organizations will adapt to these changes and thrive in the future."

Praniti Lakhwara, CIO, Apttus

"Mark Settle is not only one of the CIO thought leaders in the Valley but always generous with his time and advice. This book is another example of his many contributions to the IT community and shows why his perspective is so appreciated."

Karl Mosgofian, CIO, Gainsight

"*Truth from the Valley* is a strategic planning guide for every IT organization at the outset of the 2020s. Fundamental changes in talent acquisition and development, internal organizational processes, and technology management are occurring all around us. Every IT shop needs to take stock of these changes and determine how they will be impacted. *Truth from the Valley* will guide you through this strategic assessment and assist in developing a plan of action that is uniquely suited to your company."

Mark Grimse, Former CIO, Rambus

"Mark Settle's practical advice is based upon his personal interactions with many of Silicon Valley's leading companies. It's been of great help to me and many others. Mark has a unique perspective and a wealth of experience and knowledge to share."

Brian Hoyt, CIO, Unity Technologies

"When it comes to IT trends, Silicon Valley is often a harbinger of what's to come next for the rest of the IT industry. Learning about the talent, technology and operational practices being implemented in the Valley allows the rest of us to better prepare for the challenges we'll all be facing in the future. In *Truth from the Valley,* Mark Settle provides a clearly articulated glimpse into the changes that bleeding edge companies are making to succeed in the next decade. Mark's panoramic view of the changes sweeping through our industry are thought-provoking and scary, all at the same time!"

Martha Heller, Author of *Be the Business: CIOs in the New Era of IT* and *The CIO Paradox: Battling the Contradictions of IT Leadership*

"*Truth from the Valley* is a guided tour of the strategic challenges facing every IT organization at the outset of the 2020s. Seven-time CIO Mark Settle provides practical advice on how to convert those challenges into opportunities that will make IT teams more productive and business relevant. It's a must-read for every IT leader."

Mindy Lieberman, VP, Enterprise Systems, Peleton

"Mark Settle takes readers on a twisted, unpredictable and hilarious journey into the treacherous world of IT management in Silicon Valley. Another great CIO companion guide from Mark."

Eric Tan, CIO, Coupa

"In *Truth from the Valley* Mark Settle breaks out his crystal ball to examine the talent development, operational practices, and technology management trends that are reshaping the IT industry. Mark explores ways to leverage these trends and develop the competencies that IT teams need to remain business relevant in the 2020s. This book is an invaluable strategic planning document and a must-read for every IT leader!"

Shawn Johnson, Former IT VP, Great-West Financial

"*Truth from the Valley* provides extraordinary insight into the top challenges faced by technology leaders within the progressive, leading edge environment of Silicon Valley. IT leaders should treat this book as their personal over-the-horizon radar system, warning them about changes in industry best practices that will impact them directly during the coming decade. Mark Settle shares his real world experience and practical tips on how to build a next generation IT organization. His insights are as valuable to business leaders as they are to IT professionals."

Rodney Fullmer, CTO, Arrow Global Services

"Mark Settle always has original insights into IT industry trends. This book is no exception. *Truth from the Valley* is a must-read for IT practitioners who are transitioning, transforming or reminding themselves about how to be great leaders."

Steve Comstock, Former CIO, CBS Interactive

"Famous British economist John Maynard Keyes once said, 'the difficulty lies not so much in developing new ideas as in escaping from the old ones.' *Truth from the Valley* hits on both. It clearly identifies traditional IT practices that are becoming irrelevant in a digitally transformed world and it describes breakthrough concepts that will define success for IT teams in the coming decade."

Prakash Kota, CIO, Autodesk

TRUTH FROM THE VALLEY

TRUTH FROM THE VALLEY

VALLEY

A Practical Primer on IT Management
for the Next Decade

MARK SETTLE

First edition published in 2020

by Routledge/Bibliomotion, Inc.
52 Vanderbilt Avenue, 11th Floor New York, NY 10017
2 Park Square, Milton Park, Abingdon, Oxon OX14 4RN, UK

International Standard Book Number-13: 978-0-367-43000-9 (Hardback)

Library of Congress Cataloging-in-Publication Data

Library of Congress Cataloging-in-Publication Data
Names: Settle, Mark (Chief information officer), author.
Title: Truth from the Valley : a practical primer on IT management for the
next decade / Mark Settle.
Description: New York, NY : Routledge/Bibliomotion, 2020. | Includes
bibliographical references and index.
Identifiers: LCCN 2019045042 (print) | LCCN 2019045043 (ebook) | ISBN
9780367430009 (hardback) | ISBN 9781003001003 (ebook)
Subjects: LCSH: Information technology--Management.
Classification: LCC HD30.2 .S485 2020 (print) | LCC HD30.2 (ebook) | DDC
004.068--dc23
LC record available at https://lccn.loc.gov/2019045042
LC ebook record available at https://lccn.loc.gov/2019045043

Visit the Taylor & Francis Web site at
http://www.taylorandfrancis.com

Printed and bound in Great Britain by
TJ International Ltd, Padstow, Cornwall

This book is dedicated to my father
Marlin Young Settle
whose lifelong work ethic,
instinctive ability to befriend others,
and personal patriotism
inspired me in ways I deeply appreciate now
but failed to recognize during his lifetime.

Contents

Preface

I moved to Silicon Valley in 1999 to take a position at a large financial services firm. Although I was well aware of the Valley's reputation as the global epicenter of technology innovation, my personal experience with the local venture capital community was quite limited at the time.

During the subsequent 20 years I worked in a variety of jobs outside the Bay area but maintained my Silicon Valley home and learned more about Silicon Valley culture. I had the good fortune to meet a variety of individuals working within the high-tech industry and became an early customer and informal advisor to many startup firms.

I returned to the Valley on a full-time basis in 2016 to join a highly successful late-stage startup company. Although I had some broad perceptions about how IT was managed within the Valley, the opportunity to lead the IT team at a rapidly growing startup has provided a wealth of practical and unanticipated insights. Those insights propelled me to write this book.

My return to the Valley not only gave me firsthand experience in managing IT within a cloud-native company, it also gave me an opportunity to collaborate with peers and acquaintances who grapple with similar problems and challenges every day. The insights shared in this book are not based solely on my personal experience, but also on the lessons I've learned from my fellow travelers.

There's an old African proverb that "it takes a village to raise a child." I think it's equally true that in these modern times it takes a network to be an effective IT leader. The world is becoming too complex and the pace of business is increasing too rapidly for any leader to rely exclusively on their personal knowledge to guide their teams. Business models constantly expand or contract in response to competitive pressures. Companies are becoming increasingly reliant on suppliers, service providers, and go-to-market partners for their success. And finally, as technology leaders, we are all acutely aware of the pace of technology innovation. No leader – regardless of their intelligence, skills, dedication, or experience – can truly "go it alone" in the world we live in today.

Although the opinions and suggestions within this book are solely my own, my observations and insights have been formed through spirited conversations with many others. Special thanks to Alvina Antar, Paul Chapman, Julie Cullivan, Anil Earla, Stuart Evans, Tom Fisher, Brian Hoyt, Eric Johnson, Shawn Johnson, Yousuf Khan, Diana McKenzie, Declan Morris, Trevor Schulze, Manjit Singh, Eric Tan, Adam Franklin Wickersham, and Tony Young for sharing their trials, tribulations, and personal wisdom. Special thanks as well to my friends at Blumberg Capital, Foundation Capital, Greenspring Associates, Index Ventures, Insight Partners, Landmark Ventures, Lightspeed, Mayfield, Ridge Ventures, Sapphire Ventures, and Sutter Hill Ventures for inviting me to participate in their CIO advisory groups and exposing me to many of their portfolio companies.

Hopefully, readers of this book will have also read my prior book, *Truth from the Trenches*. *Truth from the Trenches* describes the *personal competencies* that IT leaders need to manage their teams and advance their careers. It explores the common problems and pitfalls that leaders encounter in managing money, people, innovation, and business relationships. *Truth from the Valley* describes the *organizational competencies* that leaders must instill within their teams to succeed in the next decade. It's a strategic planning document that leaders can use to improve their team's business impact through targeted investments in specific organizational capabilities.

The IT industry is entering a new age at the outset of the 2020s. Revolutionary changes in talent availability, operational practices, and technology management have occurred during the past 10 years. IT leaders need to recognize the significance of these changes and leverage them if they hope to succeed. That's ultimately the purpose of this book: to identify the secular changes that have occurred within our industry and to stimulate thinking about how to convert such changes into opportunities instead of challenges.

Personal learning can be accomplished in many different ways. As you peruse this book, I encourage you to consider the learning practices employed at a grocery store in Connecticut called Stew Leonard's. Stew Leonard's is a destination grocery that attracts visitors from well outside the local area. Stew Leonard Jr. attributes part of his success to the formation of a One Idea Club during the early stages of his store's development. Members of the Club were company employees who were given $20 and asked to make purchases at nearby stores operated by their competitors. They were specifically asked to make purchases within the same department in which they worked at Leonard's. The goal of each shopping trip was to find *one idea* that they could implement at Leonard's that would make their customers happier.

I invite all readers of this book to join my One Idea Club. If you are able to take away one idea that will enable you and your teams to become more effective, more impactful, more respected during the next decade, then this book will have accomplished its goal. Welcome to the Club!

Mark Settle
San Francisco

changes that have occurred within our industry and to stimulate thinking about how to convert such changes into opportunities instead of challenges. Personal learning can be accomplished in many different ways. As you peruse this book, I encourage you to consider the leading practices employed at a grocery store in Connecticut called Stew Leonard's. Stew Leonard's is a destination grocery that attracts visitors from well outside the local area. Stew Leonard Jr. attributes part of his success to the formation of One Idea Club during the early stages of his store's development. Members of the Club were company employees who were given $20 and asked to make purchases at nearby stores operated by their competitors. They were specifically asked to make purchases within the same departments in which they work at Stew Leonard's. The goal of each shopping trip was to find something that would implement at Leonard's that would make their customers happier.

I invite all readers of this book to join my One Idea Club. It can also take away a few ideas that will enable you and your teams to become more creative, more impactful, more empowered during the next decade, then this book will have accomplished its goal. Welcome to the Club!

Mark Serva
San Francisco

Author

Mark Settle is a seven-time Chief Information Officer with broad business experience in the enterprise software, information services, consumer products, high-tech distribution, financial services, and oil and gas industries. He has led IT teams that support the internal business operations, product development activities, and revenue-generating websites of multiple Fortune 500 companies. Mark has served as an informal advisor to many Silicon Valley startup firms and currently sits on the advisory boards of several venture capital companies. He is the author of *Truth from the Trenches: A Practical Guide to the Art of IT Management* and is a three-time *CIO 100* honoree.

Introduction

"The future is here, it's just unevenly distributed."
William Gibson,
science fiction writer

This is the foundational premise of this book. The future of IT management is here and it's being pioneered within Silicon Valley. The challenges, issues, and opportunities being faced by IT leaders in the Valley today will be confronted by their colleagues throughout the world in the future. Perhaps their colleagues won't experience a wholesale revolution in their current operational practices but many of the ways in which IT is managed in the Valley will inevitably appear elsewhere.

Why is Silicon Valley the leading laboratory for reinventing IT management? What is it about the people, processes, and culture of the Valley that make it an ideal incubator for prototyping the management practices of the future?

The rules of engagement between IT and functional business leaders have been fundamentally rewritten in the Valley. Functional groups routinely select, buy, implement, and maintain the Software as a Service (SaaS) applications required to support their internal business processes with minimal, if any, IT involvement. Functional groups have funding within their budgets to purchase software subscriptions and professional consulting services directly from SaaS vendors. In extreme but not uncommon cases, functional teams may even procure subscriptions or services without IT's knowledge. More commonly, IT is included in the application procurement process at the eleventh hour to address security and data management concerns after the winning vendor has already been selected.

IT's role in maintaining and administering SaaS applications has been similarly diminished. Functional teams routinely configure such applications, administer user access privileges, manage data quality, and generate reports with little, if any, IT assistance. IT's residual responsibilities are primarily concerned with synchronizing data across multiple applications and enforcing information security policies.

IT infrastructure management has become democratized as well. Bring Your Own Device (BYOD) smartphone policies have been implemented in many companies, in practice if not in principle. IT departments are retreating from the administration of company-purchased devices and are enforcing security controls through various mobile device management (MDM) tools. As MDM technology has become more sophisticated and reliance on cloud-based SaaS applications has grown, the use of virtual private network (VPN) technology to secure mobile access to key systems has waned.

Compute and storage technologies are increasingly being consumed by employees on a self-service basis through the use of Amazon Web Services, Microsoft Azure, or Google Cloud. Software development and application support teams can easily provision the resources they need to support their day-to-day activities with minimal IT intervention. As in the case of SaaS management, many of these teams have their own budgets and procure cloud-based resources directly.

DevOps practices – which are wildly popular within the Valley – further disintermediate IT from its conventional responsibilities for monitoring and maintaining business-critical systems. Development teams have their own monitoring tools and their own practices for managing a wide variety of production support issues. Some have established formal roles for dedicated Site Reliability Engineers, usurping the support responsibilities of many conventional IT infrastructure teams.

It's ironic – and somewhat scary – that after years of complaining about "shadow IT" activities occurring within various functional groups the roles and responsibilities of IT teams in the Valley have diminished to a point at which they've become shadows of their former selves. The residual responsibilities of most Valley IT shops focus on integrating data among various SaaS applications, managing networks and endpoint security, administering personal productivity tools such as Slack and Box, maintaining the enterprise data warehouse, and enforcing security policies. For most employees, these are largely shadow activities that don't impact their daily lives unless something goes wrong. As a result, conventional IT groups have become shadow organizations in many cloud-native Valley companies.

The IT management culture is less hierarchical, less insular, and less risk averse. Silicon Valley is admittedly a somewhat unique environment possessing lots of brainpower, lots of investment dollars, and a small army of individuals who have successfully turned clever ideas into profitable companies. It's compact, consisting of portions of Santa Clara, San Mateo, and San Francisco counties. You can drive from one end of the Valley to the other in 90 minutes (depending on traffic!). Individuals working in all functional areas, including IT, change companies with relative ease.

The IT job market is tight and highly competitive. Talent retention is a continuous challenge. It's not uncommon for the recruiting team at a startup company to receive dozens of unsolicited resumes from potential job seekers prior to their company's initial public offering (IPO). These individuals are not applying for specific jobs. They're seeking any job that might fit their skills because they're attracted to the company's technology, leadership, and/or business potential.

The close geographic proximity of so many high-tech firms and the varied work backgrounds of Valley employees promotes a high level of casual, personal interaction. Valley workers and managers are intellectually promiscuous. They're keenly interested in business and technology concepts being promoted by newly formed startups; venture capital investments in mid-to-late-stage firms experiencing strong market traction; and the movement of key individuals among different tech companies. IT leaders routinely compare notes on the capabilities and limitations of tools being commercialized by new firms. There's a strong element of intramural competition among local leaders seeking recognition as early adopters of emerging tools and technologies.

Venture capital (VC) firms play a key but subtle role in promoting interactions among Valley IT leaders as well. They routinely host meetings, workshops, and dinners showcasing their current and prospective investments. They regularly seek feedback from local leaders regarding the business value propositions of individual startup firms.

Many Valley IT shops have become adept at being "fast followers," relying on the experiences of their colleagues at other firms to accelerate the evaluation and implementation of new technologies. That's rarely the case in larger companies with significant investments in conventional hardware and software systems. Larger IT organizations with entrenched vendors and long-standing ways of doing business are frequently wary of new technologies and new operating procedures. They have an instinctive "not invented here" reaction to new ideas and subject new practices to rigorous, time-consuming proof-of-concept evaluations and piloting exercises. It can be easily argued that Valley IT shops are smaller and less technically complicated than their larger counterparts and therefore find it easier to implement new tools and processes. However, it's equally true that many Valley shops start their evaluations of new ideas with a "why not?" attitude instead of asking "why bother?"

Valley leaders may not be inherently more curious or more highly networked than their counterparts elsewhere but they clearly benefit from the relative absence of technical debt within their organizations. Conventional IT shops in larger, more well established companies are mired in the support and maintenance of critical business systems built upon aging or obsolete technology. Tech debt is not only a huge intellectual distraction for the leaders of such organizations, but it makes their experience and insight less valuable because the debt dilemmas they face are uniquely determined by the technology decisions their companies have made in the past. Tech debt can hold the leaders of conventional IT organizations hostage, both in terms of their intellectual focus and aversion to risk. Tech debt numbs innovation by diverting investment dollars, staff time, and management creativity to the job of simply keeping current systems up and running.

Employee aspirations and expectations go far beyond simply holding a job. It's a bit of an exaggeration, but IT in the Valley is generally an avocation in addition to being an occupation for many IT professionals. All professionals want to work in successful, growing companies that will expand their skills and advance their careers. But in the Valley, they frequently seek much more. They're keenly interested in working for companies that are commercializing radically new technologies like blockchain or machine learning. Their definition of a growing company is not one that's increasing revenue at 10% per year. They're hoping to work for companies that are doubling revenues every year. Finally, they carefully examine the job pedigrees of a company's leadership team because they realize they're not simply joining a company's workforce – they're being inducted into an extended professional network and the company's leaders are the central nodes in that network.

IT professionals everywhere change jobs to advance their careers. In the Valley they simply have more options. Large successful companies like Salesforce, Google, LinkedIn, and Facebook serve as talent magnets. They attract talent from all over the country and provide advancement opportunities for existing Valley professionals. These same companies supply talent to startups, giving professionals the opportunity to work on newer technologies, assume more impactful roles in smaller organizations, expand their personal networks, and obtain equity stakes in pre-IPO firms.

Valley professionals are frequently invited to consider new job opportunities by their colleagues. Referrals typically account for 40–60% of the job candidates ultimately hired by startup companies. Candidates don't simply accept jobs because they're attracted to the scope and nature of their new role, they change jobs to work with their friends.

There was a popular parlor game in the early 2000s called Six Degrees of Bacon. It was based on the premise that anyone working in the movie industry could establish a connection to the actor Kevin Bacon through a series of five or fewer serial acquaintances. A variation of this game could easily be played within Silicon Valley. It's likely that all IT professionals can establish a connection to any tech firm in the Valley through two or fewer acquaintances. They can probably connect to a hiring manager at any firm in four!

People Change Jobs for the Funniest Reasons

As a rule, I rarely interview candidates for staff jobs within my IT organization because I don't have the technical aptitude to evaluate their credentials. Furthermore, since it's unlikely I'll work with them directly, my personal opinions regarding their collaboration skills are far less valuable than the opinions of their potential co-workers. However, I do try to spend 30 minutes with newly hired staff members to welcome them to the organization and learn about the factors that led them to accept our employment offer.

I recently had a post-hire conversation with a young woman who had come from a successful technology firm. In describing her background, it became clear that she had been very successful at her former company. She had worked on a variety of strategic business initiatives, introduced new technologies that played a critical role in expanding their business operations, and had received multiple promotions and financial rewards. She had established an enviable track record of progressive accomplishment at her prior firm so I asked her why she decided to leave.

She told me that the IT team within her former company had experienced two hiring spurts during the past 5 years. This had created an "older crowd" of staff members with 4–5 years of tenure and a "newer crowd" with tenures of 2 years or less. The older crowd kept having to explain to the new crowd why their IT systems and processes were configured in certain ways. The new crowd had new ideas and the older crowd's explanations were becoming increasingly defensive.

One day this woman realized that she was behaving and being treated like a member of the old crowd! She had never thought of herself as a defender of the status quo in any of her prior positions and the realization that she had unwittingly become a member of the old crowd in her former company motivated her to seek employment elsewhere. She was passionately interested in building something new rather than defending something old and felt that our company was new enough and young enough that she could really make a difference. That was the single most important reason why she chose to join us.

These changes are not just happening in Silicon Valley. They're breaking out in Silicon Alley (New York City), Silicon Beach (Los Angeles), Silicon Forest (Portland), and Silicon Prairie (Dallas-Fort Worth). They're also occurring in smaller, less well publicized communities, such as Indianapolis, Salt Lake City, and Nashville. They're occurring around the edges of large, well established enterprises with deeply entrenched ways of doing business. IT leaders across the United States are confronting some or all of these changes to one degree or another. They're inescapable and will only become more pervasive with the passage of time.

A new IT operating model is needed for leaders and their teams to be successful in the next decade. This model needs to do more than simply respond to the changes described above. It needs to leverage them in ways that will make IT organizations more efficient, more effective, and more impactful in the future.

What's the best way to think about a new operating model for IT? *Business models* describe how companies capture, create, and deliver value through interactions with their suppliers, partners, and customers. *Operating models* focus on the transactional processes that enable the delivery of profitable products and services. They prescribe the policies and procedures employed to implement a company's business model and realize its intended value proposition.

An *IT operating model* is an integral part of every company's overall operating model. It delivers the IT capabilities that are needed to enforce the policies and execute the practices that are required to achieve a company's value proposition. Successful IT leaders are obsessively focused on delivering value to both internal and external stakeholders because they recognize the critical linkage between their operational capabilities and the success of their company's business model.

People, process, and technology are the foundational pillars of any organizational operating model. They are the time-tested axes of organizational management. They've been used for more than half a century by an army of consultants to help enterprises respond and adapt to changing circumstances. This book is organized around these three pillars.

The following discussion addresses people, process, and technology in that prioritized order. This prioritization may dismay some technologists who naturally gravitate to discussions of technology as the primary source of IT's business value. Doctrinal technologists believe that technology is the wellspring of IT's value, processes are nothing more than institutionalized administrivia that must be endured, and people are interchangeable piece parts required for the maintenance of technical systems. Experienced leaders – at some point in their careers – come to the realization that exactly the opposite is true. Therefore, the following discussion will revert to the prioritization employed by the army of consultants referenced above: people first, processes second, and technology third.

Readers of this book may ask how it differs from my first book, *Truth from the Trenches*. *Truth from the Trenches* focused upon the competencies that IT leaders require for their personal success. This book focuses on organizational competencies that are required for IT teams to succeed in the next decade. It is every leader's responsibility to cultivate and institutionalize these competencies. This book is not a sequel to the first but readers of both will need to excuse some periodic plagiarism on topics that provoke the personal passion of the author!

PART I

People

"Talent wins games, teamwork and intelligence win championships."
Michael Jordan,
14-time NBA All-Star, 6-time NBA Champion

Whether we like to admit it or not, IT organizations and their leaders spend far more time trying to remedy the technical debt within their systems than remedying the talent debt within their teams. Leaders and staff members pay lip service to the importance of people in their organizations but if you truly examine the ways they use their time you'll discover their actions belie their words.

For the purposes of the following discussion, the term "talent" is used in the broadest possible sense. It refers to more than simply skills, knowledge, and experience. It includes personal aptitudes, attitudes, and abilities. It encompasses a willingness to learn; the ability to work collaboratively with others; the willingness to share; the ability to challenge; an aptitude for assessing risk and embracing change; a sense of personal dedication and accountability; an innate curiosity about how the business works; and much more. Talent on a professional sports team is more than simply performing at your assigned position. It's making others around you better as well. In sports and business, we call that teamwork.

Every organization has talent gaps and people issues. Some have festered for months or years without being addressed in any meaningful fashion. In contrast, production issues occurring within technical systems are routinely documented and tracked. They're ranked in importance as Priority 0, Priority 1, and Priority 2 concerns. They're addressed with commensurate levels of management attention and urgency. Severe production issues receive the highest level of management attention and are monitored obsessively until they're resolved.

It's curious and revealing that no similar system exists for managing talent-related issues on a routine basis within IT organizations. Although employee performance ratings are supplied by managers, corrective action plans discussed during normal review cycles are primarily viewed as an employee's responsibility. In many cases, such plans are never explicitly developed or documented. In other cases, they're developed but rarely discussed until the next review cycle. When performance improvement plans are more formally documented, their intent is frequently to establish a rationale for eventual termination, not to correct the underlying issues that are responsible for an employee's performance deficiencies. The resolution of employee performance issues clearly lacks the level of accountability that is routinely applied to production support issues.

Why is there such wholesale avoidance of people-related issues? The first and most obvious answer is that most of today's IT managers started their careers as individual technical contributors. They never aspired to people management roles and consequently they never went out of their way to develop people management skills. They do, however, like to be promoted and consequently they find themselves assuming people management responsibilities in exchange for more pay and grander titles.

Most first-time managers with technical backgrounds abhor annual performance reviews, are clueless about how to respond to issues identified in employee engagement surveys, and struggle to complete succession planning exercises because they can't imagine that any member of their current team is potentially capable of performing their jobs. Most are notoriously inept people managers which is a direct reflection of the importance they place on the development of their team members or the depths of their personal insecurity in dealing with people-related issues.

A second factor that contributes to the avoidance of talent issues is the difficulty of effecting change. IT may not have the funds needed to expand

the skill base of the organization. The existing workforce may be complacent and largely satisfied with the status quo. They may resist the introduction of new technologies or work practices through passive–aggressive
behaviors in which they publicly extol the value of such initiatives while
doing everything they can to undermine them. The HR processes required
to put employees on notice for performance deficiencies may be cumber-
some, time-consuming, and onerous. Furthermore, managers may be reluctant to trigger the anxieties of other staff members by dealing forcefully
with performance issues. Finally, hiring managers may be genuinely concerned that they won't be able to find or recruit the human resources they
need to fill open positions. Consequently, they resign themselves to the
status quo as well.

A third factor that distracts organizations from confronting their talent
debt issues is the "superhero culture" that pervades so many IT shops. No
matter how unreasonable the demands are from IT's business partners or
how desperate the crisis may be with an existing system, it always seems
that selected members of the IT team can muster superhuman efforts to
deliver a satisfactory solution. If every demand can be met and every crisis
resolved, then there can't really be skill or capability issues within the IT
team, right?

Individual leaders may suffer from a personal superhero complex. Leaders operating under this delusion subliminally believe they are smart enough,
insightful enough, and influential enough to solve all problems and overcome all obstacles. As these individuals progress in their careers and assume
progressively broader responsibilities they ultimately come up against the
scalability challenge. They are forced to realize – sometimes painfully – that
no matter how smart they are and no matter how hard they work, they simply don't possess the breadth of skills and reservoir of time needed to make
all the critical decisions within their organizations.

Superhero cultures are not a sustainable means of supporting the
demands of an expanding business. It's the responsibility of every IT leader
to ensure that their business colleagues are not misled by tactical successes
that disguise the strategic talent deficiencies within their organizations.

The Five Levels of CIO Consciousness

Several years ago, I was asked by a journalist how the role of the CIO had changed over time. (This is a very popular interview question. I've been asked it multiple times.) I told her that I found it difficult to separate how the role had changed from how I had changed after holding a series of CIO positions. That was clearly not the answer she was expecting.

As a first-time CIO I spent most of my time trying to prove to myself that I was capable of doing the job. I worked long hours and did my best to acquaint myself with all the activities being conducted within my organization. I thought I was being hugely helpful. Individual staff members seemed to welcome my interest at the time but with 20/20 hindsight I'm sure that many managers and technical leaders thought I was micromanaging their responsibilities. I've followed first-time CIOs in two subsequent positions. In both cases the teams I inherited were relieved to discover that I didn't devote the same attention to operational details that my predecessors exhibited. I've concluded that obsessive micromanagement is a common curse for most if not all first-time CIOs.

As a second-time CIO I focused on convincing my direct reports that I was capable of managing them and orchestrating their activities. I was continually seeking validation that they were benefiting from our team interactions. I wanted confirmation that the injection of my knowledge and insights into our team discussions was adding value and assisting them in performing their duties.

In my third incarnation as a CIO, I inserted myself into activities or decisions where I thought I had sufficient knowledge and experience to provide useful advice. I wasn't terribly discriminating in determining when or where to provide such advice. I simply thought that if I had a point of view on a particular topic, I would try to be helpful by sharing it. In retrospect, I'm sure that I participated in way too many meetings where I was subconsciously trying to prove that I was the smartest guy in the room. I try not to dwell on this level of CIO consciousness because I suspect I was insufferable.

In my fourth CIO assignment, I tried to function as the Chief Quality Control Officer of the IT organization. I chose to participate in major vendor selection decisions and in the planning of major business initiatives. I had come to realize that I didn't have the bandwidth to involve myself in all the critical decisions taking place within my organization, but I did my best to become involved in those that would have the greatest strategic impact on the IT budget or relations with our business partners. I deluded myself into believing that I could participate in these types of activities as a peer, offering my opinions for objective consideration by others. In retrospect, I'm certain that the majority of managers and staff members involved in these activities interpreted my opinions or suggestions as decisions and responded accordingly.

As my technical skills atrophied and my management responsibilities broadened, I reached the highest level of CIO consciousness. I realized that the easiest, perhaps only, way to succeed was to build really strong teams. Steve Jobs once said, "It doesn't make sense to hire smart people and then tell them what to do. We hire smart people so they can tell us what to do." I have lived by this principle in my most recent CIO roles and it has paid major dividends. This is the fifth level of CIO consciousness: building effective teams and not personally managing specific activities. Fifth-level CIOs focus on three things: obtaining resources, managing company politics, and providing performance feedback to their leadership team. Departmental leaders are responsible for managing the organization. The CIO is shaping it, guiding it, and focusing it on areas where it can have the greatest business impact.

As my career has progressed my experience has become more valuable than my technical knowledge. I'm generally pretty adept at asking all the right questions and usually pretty clueless about what constitutes the correct answers to those questions. I've also discovered that suppressing my micromanagement compulsions – whether they are overt or subliminal – unlocks the creativity and initiative of the individuals in my organization.

Too many IT organizations pay too much money to too many individuals and then either tell them what to do or place cultural boundaries on their freedom of thought, freedom of expression, and freedom of action. Jobs was right. Why pay so much money for smart, accomplished people if an organization isn't willing to truly leverage their capabilities?

It can easily be argued that talent and teamwork are the sole remaining sources of competitive business advantage within most IT organizations. SaaS applications and cloud-based infrastructure services can be readily procured at a wide variety of price points. Dozens of consulting firms are available to assist organizations in implementing best practices for Service Management, Agile Development, DevOps, Project Management, Vendor Management, etc. If technology and best-in-class operational practices are readily available to all, talent and talent management emerge as the principal means of differentiating the effectiveness of one IT organization from another.

Very few IT shops can claim they've developed proprietary processes or unique uses of technology that are a source of competitive advantage. Companies operating within the same industry are likely to employ many of the same business applications, infrastructure resources, and operational practices. Their IT teams undoubtedly possess many of the same skills. The difference in performance and business impact comes down to people and team culture. People are the catalyst that ultimately allows the combination of skills, processes, and technology in one organization to deliver business value that far exceeds the value created by comparable teams possessing the same structural capabilities.

What's required to reach the moment of enlightenment in which an organization's talent issues take center stage? It may be an event. Perhaps a competitor has accelerated the time-to-market of new products, improved customer satisfaction, or reduced operating costs through the innovative use of some new technology. Perhaps it's a merger or acquisition event that exposes the talent deficiencies within the acquirer's team when they're compared to their counterparts in the acquired company. Perhaps it's the arrival of a new CEO, CFO, or COO who simply has higher expectations regarding the role that the IT department should play in promoting the growth and profitability of their company. In most cases, however, enlightenment occurs over a longer period of time through a series of missteps and failures that expose an organization's talent deficiencies and undermine its credibility, influence, and business relevance.

Five-Stage Program for Confronting Talent Debt within Your Organization

Elisabeth Kubler-Ross was a famous psychiatrist. She chronicled the emotions that humans experience in coming to terms with severe illness in her widely acclaimed book *On Death and Dying*. While in no way meaning to diminish the significance of her observations regarding the ways that individuals cope with personal tragedy, the five emotional stages described in her book provide a useful framework for characterizing the stages of self-realization that IT organizations experience in coming to terms with their talent debt.

Kubler-Ross identified the following progressive stages of dealing with severe illness: Denial, Anger, Bargaining, Depression, and Acceptance. For the purposes of this discussion, we will modify the last step of this sequence and turn Acceptance into Action.

Denial. Organizations in denial believe that their existing talent pool is sufficient to address their near-term needs. They may acknowledge gaps in skills, breakdowns in teamwork, or the presence of underperforming team members but conclude that none of these issues is materially impacting the effectiveness of their team. Dysfunctional families display similar characteristics. They tolerate conflict and misconduct, openly neglect selected family members, and fail to hold one another accountable on such a consistent basis that they tend to treat such aberrant behaviors as perfectly normal and acceptable.

Anger. After a prolonged period of denial, organizations are forced to acknowledge lapses in performance due to a lack of talent and teamwork, usually through the failure of a major business initiative or a critical business system. Business executives ask: "How could this happen?" and "Who is to blame?" IT leaders and team members ask themselves the same questions. After years of denial this is the stage in which teams and their leaders feel victimized by circumstances they personally tolerated and perpetuated.

Bargaining. Organizations initiate a series of superficial, half-hearted initiatives at the Bargaining stage of the process to camouflage their deficiencies. These initiatives might include such things as outsourcing selected services, hiring contractors possessing much-needed skills, employing management consultants to re-engineer internal processes, etc. None of these initiatives, pursued individually or collectively, can fully address skill gaps, performance deficiencies, and teamwork lapses within the existing workforce.

Depression. At this stage it has become clear that the piecemeal solutions launched during the Bargaining stage are not addressing the systemic talent management issues within the organization. As the sheer scope and magnitude of the changes required to restructure, reskill, and restaff the organization become apparent, managers and their teams may feel overwhelmed and become demoralized. A classic response is: "They're going to have to find someone else to come in here and run this place."

Action. At the end of the journey, Action is not only possible but it's probably inevitable. The only question is whether Action will be initiated by the existing leadership team or a new regime of managerial and technical leaders. The velocity of organizational transformations accomplished by new leaders can be remarkable. IT organizations routinely establish institutional phobias that allow them to sidestep their performance issues. "We'll never use SaaS applications within this company because we can't guarantee the security of our data in the cloud." "We can operate our data centers more cost-effectively than Amazon or Microsoft." "Our customers would never perform those types of transactions on the Internet." Two years later, after a change in leadership, the same organization will have deployed over 20 SaaS applications, moved 25% of its data center workloads to the cloud, and implemented a mobile application that customers are avidly using to perform "those types of transactions." All the technologies and processes needed to accomplish these objectives were available to the prior team two years ago. However, they were unable to achieve similar results due to deficiencies in talent and teamwork. Action may be painful but it is possible if leaders rectify the talent and teamwork issues within their organizations.

A Situational Analysis of the War for Talent

The phrase "war for talent" was coined in 1997 by Steven Hankin, a management consultant at McKinsey & Company. That war has spread throughout the IT industry during the past two decades. Raids and skirmishes can be found almost everywhere but the frontlines in IT's talent wars are found in Silicon Valley.

IT talent is a renewable resource but in the short term the available talent pool is relatively finite and becoming more expensive all the time. Demand continues to outstrip supply. Schools in North America – universities, junior colleges, and vocational schools – are producing graduates with IT-related skills at a snail's pace relative to demand.

The good news is that talent has become much more geographically accessible. Modern collaboration tools make it much easier for employees and contractors to work just about anywhere. IT shops routinely employ full-time and part-time staff members located hundreds or thousands of miles from their company's brick-and-mortar offices. The bad news is that talent has become much more geographically accessible. Someone in Kansas

City with advanced data engineering or website design skills may already be employed by a company in Minneapolis and may not be available to a Kansas City employer.

Leading-edge talent is frequently immobilized by the high-tech community. Many computer science graduates from Carnegie Mellon or Georgia Tech or UT-Austin would prefer to start their careers at Google or Facebook instead of joining a traditional Fortune 500 company. Startup firms also immobilize talent by attracting seasoned IT professionals and employing them through successive stages of venture capital investments.

The county-wide unemployment rate in Santa Clara County was 2.1% in May 2019. This includes technology and non-technology jobs. The IT unemployment rate was probably less than 1% and easily zero in specific subdomains such as information security, artificial intelligence, and blockchain technology. There's a fierce war for talent in the Valley and that war is likely to intensify in other portions of the United States if it hasn't done so already. Firms elsewhere may not be locked in the war just yet, but they're likely aware of the combat being experienced by some of their competitors or other firms within their local market.

The implications of the war for talent are profound and need to be incorporated in IT's new operating model. Recruiting needs to become a perpetual activity, not just a series of event-driven initiatives triggered by staff departures or budget increases. Skills development for existing staff members needs to be taken seriously, since the supply/demand imbalance is unlikely to shift to an employer's favor anytime soon. Performance management needs to be taken equally seriously and performance bars need to be continuously raised. Talent debt is not a static condition. It increases annually as the technologies underpinning existing systems age and become obsolete. Rising performance standards are the most readily available means of ensuring that talent debt doesn't become worse.

None of the measures listed above are new ideas. Nor are they terribly profound. What has changed is their criticality. *The supply, geography, and immobilization issues described above impede every organization's access to talent. The ability to overcome these issues and replenish a firm's talent pool on a sustainable basis will become an essential survival competency of successful IT shops in the 2020s.*

The Talent Management Operating Model

Attracting Talent

Talent attraction is an exercise in brand management. Every IT organization – whether in the Valley or elsewhere – has a reputation. The leaders of IT organizations have reputations as well. These reputations can assist or undermine an organization's ability to attract talent.

Attraction is different from recruiting. Recruiting is a discussion between a company and an individual concerning a specific job. Attraction precedes recruiting and determines the willingness of an individual to pursue or entertain a recruiting conversation.

Most human organizations such as clubs, fraternities, and teams pride themselves on the diverse skills and personalities of their members. They believe their members exhibit a wide variety of behaviors and resist attempts to characterize or label their collective conduct. Nonmembers typically have very different perceptions. They usually find it easy to apply generic labels to groups and will readily refer to individual groups as being too disorganized or lazy, too political or hierarchical, or too formal and bureaucratic.

IT organizations have the same fate. IT shops are routinely referred to as being overly politicized, poorly led, or slow to innovate. Some are referred to as sweat shops because employees are expected to routinely work 60 hours a week, while others are characterized as 9-to-5 shops where you can work forever without ever breaking a sweat. Everyone is concerned about the advancement opportunities offered by a prospective employer but female and minority job candidates are particularly concerned with the composition of the IT leadership team as a means of gauging their future promotability. IT organizations commonly develop reputations about internal glass ceilings that may limit the advancement opportunities of specific individuals.

Organizational reputations are well established within the vendor community serving an IT organization. Sales representatives become intimately familiar with the personalities of the organization's managers and its internal decision-making processes. They readily share their perceptions with their other customers and with one another. Social networking tools such as Glassdoor can shape and publicize an IT shop's reputation as well.

The leaders of an IT organization, specifically CIOs and their direct reports, also develop individual reputations within their local communities. Their reputations can reinforce the stereotypical perceptions of the organization as a whole or diverge from those perceptions.

Brand reputations are the reasons that MIT undergrads choose to intern at Google in San Francisco instead of MetLife in New York. Brand is the reason that a happily employed IT professional takes a cold call from an unknown recruiter who is impressed with their LinkedIn credentials. Brand may ultimately be the reason that an individual agrees to interview with your organization or ultimately accepts your offer of employment. Brand is important.

Leaders and teams can't realistically expect to control the reputations of their organizations but they should at least be conscious of them and proactively shape them if possible. All members of the organization are sales representatives – whether they realize it or not – selling the positive and negative aspects of working at their company. Leaders at all levels of the organization bear greater responsibility for brand management because they're likely to have more public opportunities to discuss their team's current initiatives and recent accomplishments. Opportunities to speak at industry events, host industry meetings, or organize best practice discussions among local companies should not be casually dismissed. They all represent powerful means of managing brand reputation.

In the new operating model, brand management is not left wholly to chance. It's cultivated in an intentional manner. Enlightened organizations have come to realize that a flawed brand will only make recruiting harder than it already is!

Recruiting Talent

IT organizations typically manage their recruiting activities as a series of one-off events focused on opening and filling individual job requisitions. Requisitions are frequently created with little warning in response to employee departures or newly acquired funds. It's not uncommon for some requisitions – particularly those targeting skills in high demand – to remain open for months or even quarters.

There are many reasons why it's difficult to locate and recruit talent but the haphazard nature of the traditional recruiting process bears a significant part of the blame. In many cases, recruiters struggle to source an initial flow of candidates simply because they're unfamiliar with the requirements of the newly opened position. They need to translate the job requirements into a concrete set of skills, knowledge, abilities, and experience that can be used to evaluate job applicants. They typically hone these evaluation criteria through a series of screening interviews with prospective candidates before they're comfortable presenting qualified candidates to a hiring manager. Simply put, they need several weeks of practice, simply to be asking candidates the right questions and developing the intuition required to correctly interpret the candidates' answers.

Sales organizations take a much more holistic and enlightened approach to recruiting. They treat it as a continuous process. Leaders at all levels of a sales organization are assigned revenue quotas they're expected to meet on a quarterly and annual basis. The defection of even a single member of the sales team creates a risk that jeopardizes his boss's ability to achieve his assigned target. This risk cascades upward, threatening the quota attainment of other members of the organization as well. Consequently, high-performing sales teams are constantly on the lookout for new talent. Members go out of their way to monitor the success of colleagues they've worked with in the past and befriend counterparts working for other vendors who are succeeding in the local market. It's not uncommon for successful sales leaders to schedule several calls each week to informally pre-interview potential candidates to assess their potential interest in positions that are not formally open.

For better or worse, IT leaders do not have revenue performance targets and have traditionally been less inclined to cultivate external talent in advance of specific job openings. In the new operating model, they need to behave more like their peers in sales and turn recruiting into a continuous process.

It's usually not that difficult to anticipate the key skills that an IT organization will need to succeed in the future. So, instead of waiting for open headcount slots to appear as a result of attrition or incremental funding, IT leaders need to develop a working inventory of critical positions that are most likely to ensure future organizational success. Such positions may be managerial or technical in nature. In addition, leaders need to maintain an informal inventory of prospective candidates possessing the requisite credentials for such positions without regard to a candidate's current employer, availability, or geographic location. This array of prospective candidates constitutes a talent pipeline that can be tapped when specific job openings occur.

Consider this to be an ongoing succession planning process with two important differences. In this case the candidates for key positions are being sourced externally and the positions for which they're being considered don't formally exist at the present time. Conventional succession planning exercises rate internal candidates on the basis of their job performance and promotability. A continuous recruiting process rates prospective external candidate on the basis of their capabilities, accomplishments, and seducibility.

It's surprisingly easy to cultivate the attention of prospective candidates precisely because so few people do so. It's as easy as getting together for coffee, requesting feedback on a new application or collaboration tool, forwarding an email containing news or information of mutual interest, or providing introductions to other members of your professional network. These types of acts are innocuous in and of themselves but they communicate a degree of personal friendship and professional respect that will make it much easier to initiate future recruiting discussions.

There are a variety of emerging software products that can assist in managing periodic engagement with prospective candidates in a more consistent fashion. In much the same way that sales management tools can be used to monitor the frequency and extent of interactions with prospective customers, these recruiting applications can trigger automatic outbound messages to prospects or remind managers to contact prospects directly. This type of casual "drip engagement" model can have a powerful influence on an individual's willingness to consider switching jobs in the future.

A continuous talent screening process has many benefits. It enables management to hone their selection criteria for key positions without the pressure of needing to fill a specific requisition immediately. It focuses primary attention on prospective candidates who are currently employed as opposed to others who, for whatever reason, are unemployed. Longer lead times provide the opportunity to consider a broader selection of potential candidates and proactively cultivate individuals with the most highly prized credentials.

Finally, a properly managed continuous recruiting process is likely to be more successful and a more successful process is likely to attract more IT investment dollars. It's always easier to solicit incremental funds for individuals with outstanding credentials who would clearly benefit the organization and are willing to join. Success will beget success. A proven track record of attracting superior talent will be noticed by business executives and they, in turn, will be more likely to support future increases in IT staffing.

IT organizations have made extensive use of contract-to-hire arrangements to screen prospective job candidates in the past. This practice, commonly referred to as try–buy, enables a company to procure the services of a specific individual on a temporary basis with the option of extending an offer of full-time employment at the conclusion of the individual's contract assignment. This recruiting practice is well suited to an environment in which the supply of talent exceeds the demand. When supply exceeds demand there will always be a significant pool of highly qualified individuals who are forced to work on a temporary basis before they're able to find a full-time job.

Contract-to-hire practices are less effective in the new operating model in which demand exceeds supply, sometimes dramatically. This imbalance may be particularly acute in high-demand disciplines such as information security, data analytics, or cloud operations. Individuals who opt to work as contractors in this environment largely do so as a lifestyle choice or because they fail to qualify for jobs that are readily available. Those who have opted for temporary employment as a lifestyle choice are obviously not predisposed to accepting a permanent job offer. The qualifications of the remaining individuals in the contractor pool are suspect. Consequently, direct sourcing of fully employed candidates is rapidly becoming the dominant recruiting practice in the Valley. Direct sourcing practices are supported by a new generation of social networking tools that make it easier to locate talent with the desired skills, knowledge, and experience anywhere in the country.

Every candidate sourcing, screening, and selection process involves tradeoffs between time, quality, and cost. The relative importance of time, quality, and cost should be explicitly defined at the outset of every search. If a critical skill is needed for the success of a current project, time may be the overriding consideration. If the organization is attempting to build foundational competencies that are strategically important (for example, DevOps engineering skills), then quality may override all other considerations. Cost considerations come in four forms. Out-of-pocket costs may be incurred if an external recruiting firm is sourcing prospective candidates. A premium may be paid for contractors or consultants who are filling the full-time role on a temporary basis while the search is underway. Staff members contribute the cost of their labor in reviewing resumes and interviewing candidates. And finally, the organization can suffer opportunity costs by not having the expertise needed to address recurring issues or embark on new initiatives. Staff labor and organizational opportunity costs should be explicitly considered in prioritizing the importance of time, quality, and cost for every search.

Hiring managers need to be especially cautious in relying upon contractor conversions to fill open full-time roles. Although such conversions may reduce the length of a search, they may compromise the manager's ability to survey the quality of the talent pool that is potentially available. Alternatively, quality-driven searches may turn into unicorn hunts, trying to find idealized candidates that don't really exist. Managers can guard against unicorn hunts by placing absolute time limits on specific searches and reclaiming open headcount that can't be filled within the prescribed period of time. Experience has shown that hiring managers can become quite adept at completing the most difficult searches on time after they've experienced a loss of headcount.

Immature organizations waste inordinate amounts of time trying to determine if job candidates possess a complete set of the qualifications associated with a specific role. There is no such thing as a perfect candidate. Every candidate has gaps in their skills and experience relative to the role they are seeking. The purpose of the screening process is not only to detect such gaps but also to characterize their size and significance. Hiring managers are responsible for managing on-the-job performance relative to the gaps identified during the recruiting process after a candidate accepts an employment offer. Experience has shown that it's possible to manage gaps of 10% to 20% in a candidate's overall capabilities. Gaps in excess of 20% can rarely be remediated, even by the most experienced managers. Candidates with gaps in excess of 20% should never be hired.

Interviewing Tips

I've interviewed for many different positions over the course of my career. In fact, I consider myself to be a semi-professional interviewee. In many instances (maybe 30–50% of the time) it's blatantly obvious that the interviewer has given no thought whatsoever to the scope or structure of our interview conversation. It's not unusual for the interviewer to pick up a copy of my resume and peruse it for the first time during our discussion or to simply fall back on the classic introductory suggestion to "tell me a little bit about your background." Unprepared interviewers either make up random questions on the spot, talk about themselves, or fall back on a set of standard questions that they've used repeatedly in the past.

Too many interviewers fail to take the interview seriously. They act as if it's a professional courtesy or favor that they're doing for you (the interviewee) or the hiring manager. Conversations with interviewers other than the hiring manager are frequently perfunctory and largely unstructured. Some interviewers will even tell you that they don't really know why they were included on the interview schedule!

I interviewed for one CIO position several years ago in which I was asked how much I had reduced IT spending in my former position, what collaboration tools would be best suited to their work environment, whether SAP or Oracle was the superior enterprise resource planning (ERP) system, and how to move files within SharePoint. These questions were posed by a series of C-level executives and SVPs. By the end of the interview day, it was painfully apparent that this company had no real idea about the role they wanted the CIO to play and that they had never discussed the qualifications they were seeking in the successful candidate.

If talent plays such a crucial role in the success of the IT organization and incremental headcount is so hard to obtain, why do leaders and staff members treat interviews so cavalierly? Here are some guidelines to consider in planning your next interview.

Ensure Your Personal Success

The stated purpose of an interview is to determine if a candidate has the technical knowledge, work experience, and personal maturity to contribute to the goals of your organization. This is a professional and politically correct way of saying: "Will this person make *me* more successful?" Hiring managers are ultimately trying to determine whether a candidate brings capabilities to their team that will make the team, and consequently themselves, more successful. Candidates can contribute to teams in a variety of ways. They may possess technical skills, industry-specific domain knowledge, an aptitude for managing virtual teams, the ability to work effectively with business partners, financial or vendor management experience, or any of a wide variety of other capabilities that will directly contribute to a team's success. To be perfectly blunt, a hiring manager should be trying to determine whether a candidate can do things that the manager's existing team can't do at all, can't do well, or doesn't want to do in the future. The manager should also determine any candidate shortcomings that could be a potential drag on their personal success. Will the candidate require inordinate management attention? Will the candidate require time to develop capabilities in areas that complement their prior job experience before they can contribute to the team in a meaningful fashion? Will the candidate be able to work collaboratively with other team members and the team's stakeholders? After considering the pros and cons, if the candidate isn't going to make you and your team materially more successful, why would you ever hire them?

Test the Box

It's human nature to categorize people and place them in a mental box. This is a basic human coping mechanism. It's an intellectually lazy way of dealing with the information and sensory inputs we receive when meeting someone for the first time. Boxes are nothing more than frames of reference built upon preconceived notions, personal biases, and personal experience. They aren't necessarily discriminatory, but they can be. Mental boxes are responsible for the phenomenon of love at first sight.

We each have our own personal set of mental boxes and we use them to categorize job candidates. We typically place far too much reliance on preconceived associations during the interview process. Just because a candidate graduated from Stanford, worked for Facebook, has a distant relative in your home town, and is a passionate Golden State Warriors fan doesn't automatically qualify them for a position on your team.

Testing the box is an exercise in testing your preconceived notions about a candidate's capabilities based upon their past associations. Candidates are awarded the benefit of the doubt concerning the value of their prior experiences far too often simply because those experiences are associated with institutions or individuals that are familiar to the interviewer.

A more nuanced approach to testing the box is to determine what Stanford courses the candidate found most challenging, what projects at Facebook failed to achieve their initial goals or fell behind schedule, who mentored them during the early stages of their career, how they developed the skills required to be successful in the position they are seeking, etc.

Break the Script

If an interviewer asks standard questions, they will receive standard answers in return. Everyone is familiar with standard questions employed in interviews. What are your major strengths and weaknesses? What was your biggest single accomplishment in your last job? How do you deal with difficult co-workers? What are your long-term career aspirations?

It's not that these aren't important questions. But if they're posed in the standard fashion, they'll elicit standard replies. The candidate's resume should be used to ask these same questions in a more personal and insightful fashion. Use the resume or the candidate's LinkedIn profile to personalize questions about past experiences, both good and bad. Alternatively, instead of asking questions, make a provocative assertion based upon the resume information and force the candidate to validate or refute your assertion. Good interviewers intentionally try to throw the candidate off balance and make them think on their feet. It provides far greater insight into their thought processes and typically elicits more genuine information as well.

Stop Talking!

I've been in many interviews in which the interviewer talked for three-quarters of the time. It's not at all uncommon for an interviewer to talk for half of the allotted time. Interviewers like to talk about themselves partly because they like the sound of their own voice, partly because they're trying to convince the candidate to join the company, and partly because they are completely unprepared for the interview and can't think of anything else to do.

The ways in which a candidate responds to questions can provide insight into their thought processes that is as valuable as the content of their response. Does the candidate pause to organize their thoughts before replying? Are their answers concise and to the point? Do they ramble and never quite fully address the question that was originally asked?

Interviewers should gauge the effort the candidate has put into preparing for an interview. Failure to prepare raises questions regarding their work habits, analytical skills, interest in the job, and professional respect for the interviewer. Candidates need to be given sufficient time to ask their questions and should be evaluated on the perceptiveness of their queries. (In a former company I was told that the competition for their CIO position had ultimately come down to me and one other candidate. I learned the identity of the other candidate and thought he was eminently qualified for the position, perhaps more qualified than me. When I asked my new boss why he had selected me for the job, he said, "You asked better questions.")

Interviewers should always invite questions but should be wary about letting the candidate flip roles and start interviewing them. If the interviewee isn't speaking two-thirds of the time or more in response to the interviewer's comments, something's wrong – stop talking!

Personally Check References

Reference checking is the final stage of the interview process. At this stage the candidate's professional colleagues are being interviewed – not the candidate themselves. Reference checking is relegated to the final stage of the process for good reason. No one wants to waste the time or effort required to contact the references of multiple candidates, most of whom will never be hired.

Reference checking is usually a pro forma exercise performed by HR to validate information regarding past employment and obtain a formal endorsement of the candidate's capabilities from their professional colleagues. These pro forma conversations are usually conducted within 10 or 15 minutes.

Reference checking should be performed by hiring managers. The candidate's references are usually close professional colleagues who know the candidate quite well. In many cases they're personal friends. They sincerely want the candidate to succeed in the new position that he or she is being offered.

Once they are assured of a company's intent to extend an offer of employment, most references will be quite frank in providing job-related information to the candidate's new manager. They may have been guarded in discussing the candidate's shortcomings during the earlier interview process but now that a job offer is about to be extended they're typically much more willing to coach the hiring manager on the candidate's personal development needs. References are unlikely to share such information with HR representatives but are willing to do so with the candidate's new boss.

Reference checking is a valuable source of information. Everyone has personal development needs. The hiring manager can learn about these needs through reference checking and address them immediately when the candidate reports to work, or the manager can rediscover them on their own during the candidate's first 6 months on the new job. It's far easier to nip performance issues in the bud if you know what to look for!

How to Measure Recruiting Success

Newly hired employees are frequently asked to develop 30/60/90 day work plans describing the initial activities they intend to perform in their new job. These plans define specific results or products the new hire is expected to deliver during their first 3 months of employment. 30/60/90 day plans provide a useful means of keeping new hires focused on their initial work objectives and also provide a valuable means of measuring an individual's progress in their new role.

However, there's a far easier and potentially more insightful way of measuring recruiting success. If, after an individual's first 90 days on the job, their manager and co-workers feel they've been around far longer than 90 days, that's a sure sign of success. Consider, for example, an individual who joins a team on March 1st. If their teammates are surprised or shocked when they're reminded the new hire has only been employed for 90 days on June 1st, that's a clear indication that the new employee has not only contributed in a substantive fashion to the team's accomplishments but has also proven to be a good fit with the team's culture. Alternatively, if the new hire is continually being referred to as "the new guy or gal" or "still enjoying their honeymoon on the new job" during their fourth month of employment, that's a warning signal that the new employee is either not contributing, not fitting in, or both!

Developing People

The talent resource pool is limited in terms of its size and accessibility. Demand for individuals with a wide variety of IT skills continues to outstrip supply. Individual organizations may suffer from structural constraints on their ability to recruit new resources related to budget limitations, the desirability of their operating locations (or lack thereof), or their brand reputations. The only sane solution for coping with these constraints is to continually expand the skills and capabilities of existing team members.

Performance Feedback

Development can only occur when individuals receive frequent, focused, and consistent performance feedback. Annual performance reviews are a vestigial management practice that has limited utility in the workplace of the next decade. They're primarily a mechanism for adjusting pay levels in response to changing market conditions. They rarely impact the motivation, work habits, or skills of individual team members. Consequently, they have very little impact on on-the-job performance.

In the new operating model, performance reviews are renamed to reflect what they are: compensation reviews. Performance feedback becomes a continuous process. In high-performing organizations, team members actually seek feedback instead of having it administered to them.

Managers need to overcome their aversion to providing feedback and view it in a much more positive light. The following observations and suggestions may be helpful.

It's not criticism, it's altruism

Most IT organizations experience periodic convulsions involving lay-offs or wholesale reorganizations. These convulsions can be triggered by adverse business conditions, mergers or acquisitions, or simply the introduction of a new leadership team. The job security of every team member is in question when such convulsions occur. Some may lose their jobs, others may be relegated to roles that do not advance their career interests. The most altruistic step that any manager can take to prepare their team members for such convulsions is to provide performance feedback on a continuous basis. Feedback gives employees the means of improving their performance and, by implication, enhancing their job security. Imagine how you would feel as a manager if one of your team members was laid off because of a performance issue that you knowingly failed to confront and rectify.

If you don't feel altruistic, be selfish

The performance deficiencies of individual team members have a collateral impact on the performance and productivity of the entire team. Peers may need to work longer hours. Work products may need to be double checked for accuracy and completeness. Rework may be required. Relationships with stakeholders may be strained, compromising the credibility of the overall team. Managers frequently ignore these side effects. The entire team learns to compensate for the deficiencies of its members in much the same way that dysfunctional families accommodate themselves to the misbehavior of selected family members. Performance deficiencies are far more obvious – both to internal peers and external stakeholders – than most people are willing to admit. Failure to deal with such deficiencies ultimately tarnishes the reputation of the team and its managers. For purely selfish reasons, team leaders and co-workers need to provide constructive performance feedback to ensure that their own advancement opportunities aren't undermined by the shortcomings of their colleagues.

Bite-sized feedback is easier to digest

The problem with annual performance reviews is that they're perceived to be summary judgments on the business value of their recipients. They aren't suitable venues for providing the activity-based coaching needed to improve an individual's performance. Major league sport franchises such as the NFL have much more enlightened approaches to managing performance. They separate contractual negotiations (i.e. compensation reviews) from performance coaching. A franchise's General Manager provides feedback on the comparative value of individual athletes and negotiates their compensation packages accordingly. The team's coaches provide feedback on footwork, stamina, ball-handling skills, dexterity, field awareness, clock management, foul avoidance, etc. Coaching feedback is delivered in bite-sized pieces, frequently on the practice field. Bite-sized feedback is easier to assimilate and act upon, and more likely to result in tangible performance improvements.

License your feedback

Most members of an IT team engage in a wide variety of activities, ranging from operations support to sustaining engineering to business-requested initiatives. Managers can license future feedback discussions by explicitly identifying developmental opportunities associated with specific work assignments. For example, a sustaining engineering project may need to be planned and coordinated across multiple technical teams and may require the supervision of external consultants. A business-requested initiative may involve a detailed requirements analysis, the implementation of a new technology, or regular interaction with senior business executives. The various facets of these different activities may play to the strengths or weaknesses of individual team members. When such activities are initially assigned, managers should explicitly underscore the opportunity an individual will have to address a real or perceived weakness during their new assignment. This turns the assignment into a developmental opportunity and provides the manager with a license to provide targeted feedback on the developmental need that has been identified. *Activity-based feedback is a much less threatening and more constructive means of addressing performance issues than periodic performance reviews because it's topically focused, delivered in smaller increments, and presented in a business context.*

Coaching versus Managing

Too many IT managers get trapped in formal corporate processes for delivering and documenting performance feedback. They fail to recognize or appreciate the benefits that can be realized through effective coaching.

Managing has conventionally been considered to be a means of providing direction, instruction, and feedback to employees in a hierarchical fashion. Individual employees take direction from their immediate supervisor and report progress and accomplishments upward through their company's chain of command. Managers provide feedback on the timeliness, quality, and completeness of an employee's work products as well as the ways in which they perform their assignments (i.e. *what* they have done and *how* they have done it).

Coaching is a more personalized experience that's tailored to the needs and motivations of specific individuals. Good coaches function more as mentors than managers. They earn the respect and trust of their team members and are not reluctant to deliver blunt, pointed feedback – both good and bad.

Coaching is nothing more than activity-based performance feedback. Coaching occurs on a continuous basis within any major league sports team. Coaches observe team members during practice and call them out to provide on-the-spot feedback. They perform post-mortems after every game and highlight what went well and what needs improvement. Coaches and team members establish strong alignment around a common set of goals. They all want to win as many games as possible and are dedicated to developing the athletic skills and knowledge required to accomplish that goal. Good coaches have a special bond with their team members and are invested in their personal success.

Effective coaches frequently deliver feedback in a highly personalized one-on-one fashion. They intentionally employ venues or procedures that ensure the complete attention of their mentees. They may deliver feedback on the practice field out of earshot from the other team players. They may deliver it in a film room while reviewing video replays of past games. They may deliver it in their offices behind closed doors or in a corner of the cafeteria. Repetitive use of the same venue or procedure sets the stage for a meaningful feedback discussion.

Similar procedures can be employed in the business workplace. Managers may choose to conduct an immediate post-mortem meeting with selected team members following a major presentation to provide on-the-spot feedback about what went well and what could be done better next time. A manager may use an office lounge or nearby coffee shop as a feedback venue that is far removed from the distractions of an employee's desktop and surrounding co-workers. To the maximum extent possible, managers should find unique ways of staging feedback conversations with individual team members that are recognized as "coaching moments" by both manager and mentee.

Most IT organizations would benefit from far more personal coaching and far less human resource management. But true coaching can only be accomplished if managers have earned the trust and respect of their team members.

Secrets of Developing Really Talented People

There's a popular misconception that the biggest developmental challenge confronting leaders is dealing with underperformers. That's actually not true. Gaps between job expectations and job performance are usually quite apparent if someone is failing to do their job or failing to do it well. Performance shortcomings are usually obvious to everyone except the underperformers themselves!

Performance deficiencies are relatively easy to characterize and quantify. Leaders need to discuss performance gaps explicitly with individual employees, provide coaching regarding the ways in which these gaps could or should be rectified, and then step back and monitor the employee's subsequent performance. Remedial development is typically a relatively straightforward process once a manager is prepared to initiate a performance improvement conversation.

Really talented people don't always receive the developmental counseling they need precisely because they're already contributing at such a high level. What most leaders fail to realize is that such individuals could be even more impactful if their developmental needs were also explicitly discussed. Talented individuals display unique traits that must be considered in any type of feedback or coaching discussion. Really talented people don't necessarily display all the characteristics listed below but are likely to exhibit one or more.

Lack of Confidence

At some level, we all have job security concerns. Talented individuals are particularly adept at camouflaging theirs. It's been my personal experience that whenever a reduction in force is announced, the most talented individuals within my organization are among the first to ask whether their names are on the layoff list. This happens with stunning frequency. Individuals who would be the last team members to be discharged don't truly appreciate their criticality, regardless of the accolades they've received in the past. If an organization is about to experience a major restructuring or downsizing, talented individuals need to be assured early and often that they will be expected to play key roles in weathering the crisis. Other team members will take their cues from their most talented colleagues regarding the future viability of the overall team.

Overly Sensitive to Criticism

Talented individuals have received so much positive feedback throughout their careers that even the mildest forms of constructive criticism can trigger severe reactions. They may reject such criticism out of hand, vilify the messenger, or attach far more significance to such comments than was originally intended. Talented individuals may need a little more help in processing and internalizing critical feedback than other team members. More than one conversation will likely be required to ensure that such feedback has been properly interpreted.

Intellectually Promiscuous

Some talented individuals suffer from the compulsion to offer ideas, suggestions, or criticisms about a wide variety of projects and operational practices, even if they have no formal responsibility for the conduct or outcome of such activities. In many situations their feedback is very well intended. They typically possess skills, knowledge, and experience that qualify them to make such observations. Unfortunately, this compulsion distracts them from the problems or projects to which they're currently assigned. Managers need to become adept at politely but persistently curbing the intellectual promiscuity of their most talented team members and keeping them focused on those activities where their capabilities are most needed.

Value Ideation over Implementation

A related trait that frequently accompanies intellectual promiscuity is the tendency to value ideas over practical business results. Talented individuals frequently want to be respected for the creativity, perceptiveness, or novelty of their ideas. Whether such ideas are translated into operational practices that result in true business benefits can sometimes be inconsequential to them. They simply want respect for being clever. Talented individuals who suffer from implementation deficit disorder need to be counseled that ideas alone – regardless of their potential significance or theoretical value – are not a source of competitive business advantage. A relentless stream of great ideas that never get implemented is actually a sign of career failure – not career success.

Susceptible to Bouts of Depression

Talented individuals can develop a messianic complex in which they feel they're carrying an entire organization or team on their backs.

They observe the behaviors of others and conclude that their colleagues are not working as fast, as smart or as hard as they are. Under these circumstances they can become easily demoralized and depressed. In extreme cases they become passive–aggressive, marginally delivering on their commitments while paying lip service to their team's goals. Managers who detect symptoms of emotional or intellectual burnout need to find ways of getting their most talented team members back in the game, especially because other team members are likely to be influenced by the behavior of their most talented peers.

Limited Emotional IQ

The emotional IQ of talented individuals is usually inversely correlated with their technical IQ. Simply put, they are frequently oblivious to the ways in which their styles and personalities impact their co-workers. It's ironic that many talented individuals claim to welcome candid conversations and lively debates while their on-the-job behaviors suggest exactly the opposite. Managers need to find graceful ways of facilitating discussion among all the members of their teams. They may have to publicly challenge the opinions of their most talented members, simply to empower others to share their views as well. A little personal coaching can go a long way toward modifying the behavior of individuals with low emotional IQ. They're frequently shocked to learn the impact that their language and mannerisms are having on their co-workers and will modify their on-the-job behavior if such feedback is professionally delivered.

Extroverts versus Introverts

Talented individuals can have very different personalities. Some are constantly trying to prove that they're the smartest, best organized, most clairvoyant person in the room. Others are shy and introverted. They need to be invited to share their views and opinions.

Managers need to establish subtle code words and phrases for curbing the air time of talented extroverts and encouraging commentary by self-effacing introverts.

Exceptionally talented individuals are a gift to any organization. They're not simply turbocharged individual contributors. They can serve as intellectual, technical, and emotional leaders without carrying the administrative burdens associated with formal management positions. Properly coached and managed, they can have a significant impact on their teams, materially enhancing both the quantity and quality of the work being performed. Enlightened leaders realize that the benefits achieved by investing time in maximizing the performance of their most talented team members far exceed the benefits they're likely to achieve by remediating the deficiencies of their underperformers.

Development Principles

Concerns regarding personal development and career advancement are pervasive across all IT organizations. Personal development frequently receives some of the lowest ratings on annual employee engagement surveys. Employees repeatedly complain that managers are not providing sufficient coaching or developmental opportunities to prepare them for broader roles in the future.

Employee development practices have been shrouded in too much mystery in the past. To many team members it appears that a select group of privileged individuals have been given specialized development opportunities while the remaining members of the team have been left to their own devices. In the new operating model, the principles used to guide employee development need to be explicit instead of implicit. Managers need to publicly articulate and employees need to consciously understand the following principles.

There are two types of development

Somehow the notion that everyone has a God-given right to invest a specific amount of time in personal development every year has crept into our collective work consciousness. There are two types of development. *Remedial development* ensures that an individual is fully equipped with the skills and knowledge they need to succeed in their current job assignment. Remedial development is not elective. It should be prescribed by managers in consultation with individual team members. Some individuals may need more remedial development opportunities than others. *Career development* equips individuals with skills and knowledge that allow them to perform at the highest possible level within their current assignments or assume additional responsibilities not commonly associated with their current roles. Career development opportunities need to be earned based upon above-average performance in the individual's current role. Why would any manager or company want to invest time or effort in developing the careers of individuals whose current performance fails to meet expectations or is merely adequate?

Career development is a two-way street

It's not management's job to divine the career interests of individual employees. Career development needs to be an ongoing conversation between managers and team members, not a one-time annual event. Employees need to discuss career development in terms of the knowledge and experience they would like to gain, not in terms of the job titles they would like to put on their business cards. Employees and managers are equally responsible for initiating career coaching conversations on a periodic basis. Such conversations are not simply management's responsibility.

Individuals should seek career advice from others who possess the experience or responsibilities they are pursuing. Individuals who are genuinely interested in advancing their careers should cultivate a small group of informal mentors and advisors that they can use to test ideas regarding their future career paths. Company-sponsored mentoring programs rapidly lose their novelty value and usually disappear over time due to a lack of sustained commitment on the part of the mentees and mentors. Informal mentoring conversations initiated on a sustained basis by career-conscious individuals have a much higher likelihood of long-term success.

There's a right way and a wrong way of keeping score on career development

All too often, employees and managers fall into the trap of equating development with training. Formal training is usually short-lived and frequently performed outside the workplace. Skills and knowledge obtained through formal training need to be incorporated into an individual's daily work habits if they are to have any impact on current or future performance. It's difficult to do this in practice. In all too many instances, formal training is simply a holiday from an individual's normal day job. Many (most?) trainees struggle to find practical applications of their recent learnings when they return to work.

The formula for measuring IT career development is actually pretty simple. The three core dimensions are technical expertise, business knowledge, and people skills. Experiences that broaden or deepen technical expertise, expose individuals to the inner workings of their companies or industries, or improve their ability to lead and influence people are career development opportunities. On-the-job experiences in any of these dimensions are valid developmental events and should count in whatever type of personal development score an individual may care to keep. Scores based upon the frequency and nature of these experiences are a far more accurate reflection of the development an individual is receiving than simply summing up the number of days they've spent in training.

Career development does not automatically result in career advancement

There are no formulas for automatic, guaranteed career advancement. Time-in-grade formulas for advancement become increasingly less relevant in a world in which average job tenure is 3–4 years. Advancement occurs when opportunity meets preparation and hard work. As Thomas Jefferson put it so eloquently: "I am a great believer in luck and I find the harder I work, the more I have of it." This is precisely why so many people change jobs: they've done their preparation and performed the hard work but their companies were simply unable to manufacture the opportunities they desired in a reasonable period of time.

Real career development entails risk

There are very few – if any – genuine developmental experiences that are wholly risk-free. Company-sponsored development programs may provide instruction about how to collaborate with co-workers, manage a project, prepare a budget, make a presentation, or deliver performance feedback. They may employ self-assessment tools to characterize an individual's behavioral tendencies. Such programs are well intended and may be enlightening but they are almost completely risk-free. Participants frequently consider them to be more of a reward for past performance than a preparation for future challenges. No one ever fails a company development program, they simply finish the program and receive a certificate documenting their participation. Real development occurs outside the classroom. Real development occurs by doing rather than theorizing. Real development is not risk-free.

By definition, a developmental opportunity places an individual in a different role with a different set of responsibilities than they've experienced in the past. If they fail to excel in the opportunity, they may wait a long time for the next opportunity to appear. Too many employees want it both ways. They want the opportunity to work on activities or duties that fall outside the scope of their current responsibilities but they also want the assurance that they can return to their current position regardless of the outcome of such special assignments. In reality, there are few risk-free developmental opportunities. Development entails risk and employees need to consciously accept such risks, both mentally and emotionally.

Reskilling Challenges

Reskilling is the obvious solution to reducing talent debt in situations where new or additional resources are difficult to obtain. In my personal experience reskilling is easy to plan in principle but difficult to achieve in practice.

Reskilling typically occurs through some combination of training and mentoring, typically performed by external consultants who possess the new skills and knowledge required by an organization. Consultants are useful in introducing concepts and establishing frameworks. However, the devil involved in implementing new skills and processes is in the detail. Once the consultants leave or scale back their participation, team members need to develop detailed specifications for the terminology, naming conventions, documentation, review procedures, quality controls, etc. that will be required to put the new skills and processes into everyday practice. The transition from concept to practice can be painful and can take much longer than originally planned.

At the conclusion of a formal reskilling program, roughly 20% of the reskilled team really "get it." They are fully competent practitioners. Some might even have evolved into super users who can instruct and mentor others. Another 20% of the team members have barely attained minimum levels of competency. They are slow learners and are likely quite frustrated by the need to develop new capabilities. The remaining 60% have achieved adequate levels of competency but if a new position opened on the team the manager would most likely seek external job candidates who possess higher degrees of proficiency than the adequately reskilled performers. Put another way, roughly 80% of the reskilled team have failed to achieve the level of competency that managers would seek in replacing an existing team member.

There's no easy solution to this problem. Slow learners may become so frustrated that they seek job opportunities elsewhere. Adequate performers may become complacent and stop trying to achieve higher degrees of proficiency. For all of these reasons, reskilling programs almost always drag on much longer than originally planned and fail to achieve their initial objectives.

The best, perhaps only, way of accelerating reskilling initiatives is to seed the team with new members possessing in-depth knowledge of the required skills and practical experience in using them elsewhere. These individuals will shorten the transition from the consultant-led phase of the initiative to its ultimate self-led conclusion. Their knowledge and experience will pay huge dividends in translating conceptual frameworks into usable everyday procedures. Failure to seed the team with external practitioners may actually reinforce the complacency of the adequately skilled performers. They may conclude that the reskilling program will continue indefinitely until all team members have become fully proficient. The budget established for any reskilling initiative should include funds for external assistance *and* new team members.

A classic reskilling challenge confronted by almost every conventional IT shop is the adaptation of traditional data center management skills to cloud operations. In principle one could argue that the same infrastructure components – servers, storage devices, and networks – are present in both environments irrespective of whether they're physically situated within a company's data center or in the cloud. However, it's equally easy to argue that the two environments are completely different from a provisioning, monitoring, security, support, and cost-management perspective. An existing data center management team could receive formal training in cloud operations and be coached for a period of time by cloud experts. They'd subsequently be left to their own devices and learn the nuances of cloud operations through on-the-job experience (also known as the trial-and-error method!). Alternatively, the existing team could receive training and coaching and also be seeded with individuals possessing 5+ years of practical experience managing AWS, Azure, or Google Cloud operations. Which reskilling scenario would you choose?

Valley startup companies seeking to offer global cloud-based services routinely attempt to hire individuals from much larger firms such as Facebook, Uber, and LinkedIn who have managed cloud-based infrastructures at much larger scales. They've learned that it's dangerous, costly, and foolhardy to simply rely on the incremental expansion of their internal operational skills to address the scale and complexity issues created by an explosive growth in customer demands.

Development Challenges in Rapidly Growing Companies

Rapidly growing companies are commonly regarded as highly opportunistic environments in which hard work and progressive accomplishment will naturally result in career advancement. While that's true up to a point, employees working in such firms frequently discover that their personal development is stymied by two major obstacles.

The primary development challenge in rapidly growing firms is to convert newly hired employees into fully productive team members as quickly as possible. Managers and existing team members devote significant time and attention to accomplishing this goal. It's in their collective best interests to ramp new hires to full productivity as quickly as possible simply to cope with the expanding operational demands of their growing business. Many functional departments develop explicit checklists defining the activities and performance metrics that new hires must complete or achieve before they're considered to be fully productive. Some departments employ a "buddy system" in which an existing staff member is formally designated as a new hire's coach or mentor during their first 60–90 days on the job.

Comparatively speaking, the development of existing staff members is a secondary priority. Smaller, rapidly growing firms lack the formal development programs of their larger counterparts and there's less of a management commitment to set aside time and funds for purely developmental activities. Learning-by-doing is considered to be the primary means of personal development in such firms. Experiential learning is typically more effective than classroom learning but may be less beneficial in rapidly growing firms if the professional experience of supervisors and team leaders is only marginally greater than that of the team members themselves.

Employees in smaller firms tend to assume multiple roles and perform a wide variety of duties. The arrival of new employees presents an opportunity to shed or de-emphasize duties that are only tangentially related to their primary roles or involve skills that are unrelated to their long-term career ambitions. Unfortunately, in many cases, existing employees defend their right to retain all their current responsibilities and add new ones that are necessitated by the growth in business operations. We generally think about development in terms of learning to do new things. It's equally developmental to stop doing old things that are a poor use of existing skills or unrelated to an individual's preferred career path. Employees in smaller firms that bemoan the lack of personal development should seek opportunities to deepen specific skills or develop new ones by selectively shedding some of their existing responsibilities as new employees arrive.

The second obstacle encountered by employees in high-growth firms is the constant escalation of advancement criteria. As firms expand the number of employees, work locations, suppliers, and customers involved in daily operations, the qualifications associated with key jobs – especially management jobs – expand accordingly. The criteria employed to select the leader of a Service Desk team supporting 5,000 global employees will differ significantly from those used to select the leader of a team supporting 500 employees located exclusively in North America. Similarly, the best Business Systems Analyst within the application support team of an 800-person company may not qualify to be one of the founding members of the Enterprise Architecture team that's created when that company has grown to 3,000 full-time employees (FTEs).

All companies talk about continually raising their performance standards but in rapidly growing firms the escalation in performance expectations and job qualifications is tangible. It's publicly validated with every hiring decision in which an internal candidate with a proven track record is passed over in favor of an external candidate possessing superior skills, knowledge, or experience. In high-growth companies these types of staffing decisions are made frequently and visibly throughout all functional departments. Employees who have performed admirably in the past and have received praise and rewards for their accomplishments are understandably dismayed when a job they were seeking is filled by an external candidate. They commonly conclude that the rules of the game somehow changed while they were mastering the skills that presumably would qualify them for the job. They're right. The qualifications did change.

As discussed elsewhere in this book, development is an intensely personal undertaking. It needs to be planned and directed by individual employees in ways that address their unique needs and aspirations. Managers, mentors, advisors, and co-workers can assist in formulating and implementing development plans but individuals retain primary responsibility for acquiring the skills and knowledge they need to advance their careers. Formal development programs are effective in introducing concepts or methodologies but on-the-job experience remains the best and most effective means of acquiring practical skills and knowledge. Although formal developmental resources in high-growth firms are typically quite scarce, there's usually an overabundance of experiential learning opportunities available to enterprising employees who proactively seek developmental challenges.

Developmental Conversations

As discussed above, individuals and their managers need to conduct three very different types of conversations regarding performance and development. The frequency and timing of these conversations will vary. One focuses on compensation. Another focuses on performance. And the third focuses on career development. Performance conversations should occur the most frequently. Compensation conversations are most likely to occur annually. And developmental conversations should occur periodically, maybe two to four times per year.

It's dangerous to mix any two or all three of these topics in a single discussion. It's a disservice to the importance of all three. HR may force managers to link the annual compensation conversation with a discussion of job performance but if managers have been providing regular performance feedback throughout the year that portion of the compensation review will be short and uneventful.

It's also inadvisable to discuss career development during routine one-on-one meetings between managers and employees. Career development is far too important a topic to be added to the agendas of routine one-on-one meetings simply for the sake of convenience. That's not to imply that career coaching sessions necessarily need to be long. They simply need to have a singular focus. Personal development is far too important to most individuals to be randomly brought up during the last 10 minutes of a regularly scheduled status review.

It may also be dangerous to provide critical performance feedback during routinely scheduled one-on-one meetings. Such feedback may be too fleeting in nature. Individuals may choose to assume that a manager's critical comments refer to the progress of a project or activity, not to their personal performance. Bite-sized feedback is easy to assimilate but it's also easy to ignore during a weekly or biweekly review of multiple activities. Routine one-on-one meetings may not be an effective vehicle for delivering critical feedback and ensuring that individuals are taking such feedback to heart.

Developing First Line Managers

First line management is one of the most challenging, most educational, and potentially most rewarding jobs in any organization. It's certainly a developmental experience! Properly trained first line managers are force multipliers, enabling their teams to accomplish more and achieve results that far exceed the collective value of the individual work products produced by each team member. Armies win or lose battles through a series of small unit actions. The same is true of IT organizations. First line managers are IT's small unit commanders.

First line managers have typically been promoted from the technical ranks and frequently manage the pay and performance of former peers. Furthermore, they usually continue to function as individual contributors, performing certain aspects of their former positions. In sports parlance they are player–coaches, expected to spend time on the playing field themselves while directing the activities of their team members.

By definition, second line managers have been through similar experiences. They need to devote special attention to assisting first line managers in adjusting to their new responsibilities. First line managers will inevitably need the following counseling.

Stop doing your old job

First line managers may still partly function as technical contributors but they're now responsible for the work products of their entire team.

They can't simply be expected to do everything they were doing in the past and function as the team's manager at the same time. They need to determine the specific aspects of their former responsibilities that should be delegated to others and develop an explicit transition plan to ensure that such delegation occurs.

Reset Personal Relationships with Your Team Members

Whether they like to admit it or not, first line managers are not just another member of their teams. They need to establish a different type of relationship with their former peers. This doesn't mean that they need to terminate friendships but it does mean that they need to consciously avoid playing favorites in awarding work assignments, granting workplace privileges and determining pay increases. The mere appearance of favoritism may undermine the manager's credibility, irrespective of whether it's happening intentionally or unintentionally.

Start Managing Stakeholder Relationships

Every team has external stakeholders. They may depend upon the work products or services of other groups to complete their assigned tasks. Other groups may depend upon them in a similar fashion. Relationships with senior IT managers and business partners play a critical role in determining the success of almost every IT team. Although there may be multiple points of contact between team members and external stakeholders, the first line manager is ultimately responsible for orchestrating these interactions and ensuring that they remain harmonious and productive. In most cases, a first line manager's prior experience in managing external stakeholder relationships has been quite limited. In their new role, stakeholder management needs to become a daily or weekly priority.

Eliminating Talent Debt

Talent debt is a broader organizational phenomenon that includes but is not limited to performance deficiencies. In principle, performance deficiencies can be rectified through feedback, coaching, and training. Talent debt is a structural problem reflecting a more pervasive lack of the skills, knowledge, aptitudes, or experiences that are critical to the future success of the organization.

In practice, performance deficiencies can become so severe or persistent that they terminally undermine the effectiveness of an individual or serve as a drag on the productivity of an entire team. Remedial development can close gaps between job requirements and job performance that are in the range of 10–20%. Wider gaps usually indicate that an individual is incapable of performing their assigned duties satisfactorily. It's extremely unlikely that wide, persistent performance gaps can be rectified through coaching and training. Chronic underperformers contribute to talent debt because they're occupying approved positions that could be materially upgraded or repurposed with more productive individuals.

IT leaders talk about eliminating technical debt all the time. A fixed portion of every biweekly scrum cycle may be reserved for technical debt reduction. It may be featured in annual budget discussions as a major problem requiring immediate investment. But for obvious reasons leaders are uncomfortable about discussing steps that are required to eliminate talent debt.

The plain truth is that IT organizations outgrow the talents of selected staff members in the same way they outgrow the capabilities of their legacy systems and infrastructure. Reskilling is one approach to retiring talent debt and is discussed in an accompanying sidebar panel. Reskilling programs are never completely successful and organizations inevitably become increasingly less dependent upon the skills and capabilities of specific individuals.

Chronic talent deficiencies are easier to camouflage in larger, slowly growing organizations than in smaller, fast growing ones. Large organizations have two instinctive reactions to talent debt. They routinely overlook the talent deficiencies of existing team members and request additional headcount to hire individuals possessing the skills and abilities that they need. Alternatively, they may choose to rotate individuals with aging skills into new roles elsewhere in the organization, hoping that such individuals can develop skills on their new jobs that will make them more productive. In both cases, the organization is still accumulating or tolerating talent debt. It's just employing convenient ways of masking it.

Talent deficiencies in rapidly growing companies become glaringly obvious in short periods of time. The skills of individuals in key roles simply can't satisfy the demands posed by expanding business operations. Personal skills and capabilities that were stage-appropriate within a 1,000-person company based in Kansas City are unlikely to support the needs of a 5,000-person company with major operating centers in Kansas City, Houston, Raleigh, and Denver. Requirements for in-depth technical expertise and broad industry experience grow as a company matures. Individuals who were formerly praised as being "jacks of all trades" find their duties (and importance) being eroded by new recruits who are masters of specialized skills that have become critically important to the continued growth and future success of their companies.

A former Xerox CEO once famously said: "You either have to change people or change people." If current and near-term talent needs outstrip the capabilities of an existing team, then existing team members need to be terminated and replaced. Involuntary termination is treated as a communicable disease or form of social ostracism in most companies. Very few leaders seek credit or recognition for firing people. But in fact, involuntary terminations can be equally if not more impactful in improving team performance than simply hiring additional team members.

Involuntary terminations triggered by chronic performance issues or talent deficiencies can have several beneficial effects. They reinforce accountability standards throughout the organization. Involuntary terminations clearly indicate that chronic performance failures of a substantive nature will not be tolerated. They can also send a clear message to the team that the adoption of a new technology or operational practice is an organizational imperative. Individuals who are not capable or willing to support such initiatives are no longer needed.

Involuntary terminations also relieve burdens that have been explicitly or implicitly imposed on other team members. Others may have compensated for a co-worker's deficiencies by assisting them in developing their work plans, coordinating their activities, double checking their work, or correcting their mistakes. It's demoralizing to be totally responsible for one own's work and partially responsible for someone else's as well.

Specific individuals may simply be difficult to work with and complicate the ways in which assignments are structured and work is performed. In extreme situations two or more individuals on the same team may simply refuse to work with one another on specific assignments. Conflicting interpersonal relationships can have a toxic effect on team productivity. Eliminating the principal sources of these conflicts can provide a tremendous boost to team harmony and materially improve a team's work capacity.

It's not uncommon for team members to privately congratulate leaders for terminating underperforming or disruptive individuals because such terminations relieve the burdens they've assumed to compensate for the skill deficiencies or on-the-job behaviors of their former co-workers. In many cases, team members are not fully aware of the effort being devoted to compensating for a teammate's substandard skills and aberrant behaviors until that individual has departed and the burdens have been lifted.

Involuntary terminations – if properly managed – can boost the credibility of a team's managers as well as the credibility of the overall team. Team members are frequently incredulous that blatant performance or talent deficiencies have been tolerated by management for excessive periods of time. Dealing with such issues aggressively and professionally can pay big dividends in terms of a team's external reputation and the internal respect accorded its leaders. Departing team members need to be treated respectfully. While the causes of their termination may not be publicly announced, they will become well known through informal communication channels (if not already glaringly obvious to their former co-workers). Management justifications for involuntary terminations need to be stated in business terms and based upon explicit deficiencies in an individual's ability to contribute to current business needs.

Paradoxically, the biggest beneficiaries of involuntary terminations may be the terminated employees themselves. Outward appearances to the contrary, many underperformers realize that their contributions are not valued or respected by their peers. Alternatively, they may be frustrated because their prized skills and prior experiences are not being leveraged in their current positions. Terminated employees rarely thank their former managers for being involuntarily discharged but many would grudgingly admit that their departures led to better and more rewarding opportunities elsewhere. If they are truly self-reflective, they may rectify the issues that triggered their termination and be much more successful in subsequent positions. Termination may be the most developmental experience an individual can have.

Involuntary terminations can be disruptive. They frequently trigger unintended job security concerns among other team members. For this reason leaders are habitually reluctant to terminate employees. Leaders tend to give under-contributing individuals a second chance to improve their performance. There's nothing inherently wrong about providing individuals with second chances – it's just that second chances frequently lead to third chances which lead to fourth chances, etc. A fellow CIO once told me that he had developed a no-third-chance rule. He observed that underperforming individuals frequently improved their performance when they received persistent coaching and close supervision (i.e. a second chance). However, their performance issues frequently reappeared once management attention was diverted elsewhere (i.e. after their performance problems had apparently been resolved). In his experience, repetitive waves of performance coaching involved far too much effort with far too little chance of long-term success. Therefore, he established a no-third-chance rule: if performance issues reappeared after a second chance program of personal coaching and close supervision, he initiated the termination process immediately.

Leaders may fail to deal forcefully with talent debt because they're concerned they won't be able to replace terminated team members. A common refrain is: "John is only 50% effective but if I terminate him, I'll lose the headcount and then the one-half FTE that I currently have will go to zero." This concern can be partially addressed by more frequent and transparent staffing reviews. Managers within the IT organization should meet periodically to reach agreement – or at least obtain insight – into the staffing goals that are being pursued for the success of the overall organization. The financial performance of most IT organizations is reviewed on a quarterly basis and shared with the entire management team. A similar review should take place regarding staffing needs and staffing priorities over the next 2 to 4 quarters. Individual managers may not necessarily agree with these priorities but at least they'll be aware and won't be blindsided if one of their open headcounts is repurposed elsewhere. The sharing of such information won't necessarily eliminate the "bird in the hand" perceptions of managers supervising underperforming employees but it should provide more insight into their ability to replace headcount lost through attrition or involuntary termination.

Involuntary terminations are disruptive when they occur sporadically for no apparent reason. Terminations that are triggered by failures to meet explicitly defined performance expectations are accepted and understood, even if they're unpopular. Involuntary terminations are a regular occurrence within sales organizations. Sales representatives who are consistently unable to achieve their assigned revenue targets will be invited to leave the organization. These ground rules are well known and accepted by all members of the sales team. Individual terminations that trigger widespread anxieties in an IT organization may be an indication that the team's performance standards have not been explicitly defined or are not sufficiently understood.

In the new operating model, involuntary terminations are a consequence of higher and more consistently applied performance standards as well as a recognition that new skills and capabilities will continually be needed to ensure organizational success. In the increasingly competitive job market, most terminated individuals will rapidly realize that they have marketable skills and may simply have been working for the wrong company at the wrong time. The skills that were poorly suited to the needs of their former employer may be highly prized by many other companies.

Talent Debt Management Is a Lot Like IT Cost Management

Talent debt can be reduced in several ways. Talent deficiencies can be rectified through recruiting, performance feedback, on-the-job development, and reskilling initiatives as discussed elsewhere in this book. Aggressive automation campaigns may eliminate repetitive work and create opportunities to fill talent gaps by repurposing existing team members. Outsourcing the support of aging systems or employing contractors to perform routine tasks may create similar opportunities to redirect the efforts of existing staff members. Finally, involuntary terminations may be unavoidable if individuals are consistently unable to meet performance expectations or develop necessary new skills.

Unfortunately, the mechanisms referenced above are invoked all too frequently on a tactical basis in response to a specific need or event. New initiatives may require wholly new skills, forcing leaders to recruit new team members or repurpose old ones. Budget pressures may force organizations to outsource activities they formerly considered to be core competencies. Adverse business conditions may trigger wholesale layoffs that force leaders to eliminate underperforming or non-essential personnel. In all of these instances, talent debt remediation is performed on a tactical basis in the absence of a long-term plan or strategic framework. It's ironic that IT leaders manage the cost structures of their organizations on a strategic basis with formal targets in mind but fail to manage the talent structure of their teams in a similar manner.

In one of my past companies I worked for a CEO who had been a divisional manager at a major U.S. appliance manufacturer. His division made clothes washing and drying machines for home use. The American market for washers and dryers was being invaded by Japanese firms at the time. The Japanese manufacturers were consistently able to undercut his business with their pricing practices. They were consistently able to sell their products at a lower retail price and still make a profit.

Continuous cost management became a strategic imperative within his division. It was essential to their survival. All departments – from product design to supply chain to manufacturing to sales and back office – were dedicated to achieving an annual reduction of 10% in the retail cost of their products *every year for the foreseeable future*. Designs were altered to reduce the number of components within each machine. Plastic components were substituted for metal ones. Suppliers were required to preassemble certain components prior to delivery to simplify the manufacturing process. A variety of other steps were taken as well. They eventually succeeded in not only meeting but beating the retail prices of their Japanese competitors while achieving their profit targets.

Although few IT leaders are asked to reduce the size of their teams by 10% every year, almost all are challenged to find ways of continually reducing the cost of ongoing operations. Most establish explicit targets regarding the overall organizational cost structure they are trying to achieve and they use these structural targets to guide their tactical spending decisions. This never-ending process of pruning funds from one portion of the organization and redeploying them elsewhere provides an apt analogy for the way in which talent resources should be managed as well. Human resources being employed to support legacy systems and practices need to be continuously reduced or retired to ensure that new resources can be internally developed or externally acquired.

In the new operating model, an IT organization's talent resources are being continually re-engineered to address the evolving demands of its business partners. IT leaders need to proactively develop a target talent framework for their organizations and then employ all the mechanisms referenced above to reduce or retire existing resources that are not part of that framework. If leaders are not able to construct such a framework based upon their understanding of future business needs, then they've got a bigger problem. They're not spending enough time with their business partners!

Teamwork

This discussion of people and talent management opened with Michael Jordan's observation that teamwork is required to win championships – talent alone is not enough. Teamwork is so easy to describe in principle, so easy to recognize in practice but so difficult to develop and institutionalize. Teamwork is more than the glue that holds a collection of individuals together and coordinates their activities. It's the goal alignment and interpersonal chemistry that enables a group of individuals to accomplish far more than the simple sum of their independent work efforts.

The term "teamwork" is used far too casually to characterize the collective work efforts of many different types of groups. (It's probably used too casually in this book as well!) The world of professional sports is replete with teams whose members display very little teamwork. The same is true in the workplace. Many if not most workplace teams operate as working groups whose interactions are coordinated by a team leader, chairperson, or project manager. Working group members are primarily committed to the completion of their individual tasks, not necessarily the achievement of the group's collective business goals.

Teamwork is too complex of a topic to be explored in depth here but several of the key ingredients required for successful teamwork are readily apparent. Common alignment around a specific set of goals is essential. These goals need to be interpretable by all team members time-based as well. Teams should be continually reminded of the business significance of the goals they are trying to achieve.

Teamwork can only be achieved if team members are able to communicate with one another openly, professionally, and constructively. Members of high-performing teams explicitly hold one another accountable. This is what Steve Kerr, the Golden State Warriors coach, was referring to when he said that "stuff went on" in the Warriors locker room during their championship 2017–2018 NBA season. Members of successful teams are willing to challenge one another. Accountability doesn't necessarily foster harmony, but it does reinforce the commitment to common goals. Disagreement over how to achieve those goals can and should be discussed openly, even if emotions creep into such conversations. Emotions need to be channeled into a reaffirmation of the importance of a team's goals and not allowed to damage interpersonal relationships. Frayed relationships cannot be tolerated on high-performing teams. They're toxic. They must be addressed and remediated immediately by the team leader and team members.

The final defining characteristic of high-performing teams is selflessness. Members proactively seek ways of enhancing the contributions and impact of their peers. They instinctively have one another's backs when things go wrong. There are so many behavioral boundaries in the workplace that impede teamwork. If members of a specific team are convinced they're being unconditionally supported instead of being judged and criticized by their peers, they can accomplish great things. At the risk of over-indulging in sports analogies, informed observers of the Warriors 2017–2018 championship season would all agree that the selflessness of the team's members was a critical factor in their ultimate success.

Organizational skills, a sense of urgency, fiscal common sense, and stakeholder management are all secondary contributing factors to the success of high-performing teams. At their core, they are aligned around common goals, able to communicate with one another forcefully and respectfully, and have each other's backs.

It's easy to brag about teamwork when business commitments are routinely being met. True teamwork is tested and proven under stressful conditions. In IT those types of conditions occur when business-critical systems have been disrupted or compromised, or when a myriad of complex activities must be completed to meet a major project due date. Stressful IT situations are not unique to Silicon Valley. They occur everywhere. It's been my experience that effective IT teamwork under stressful conditions rarely escapes the notice of a company's senior executives. It's usually commended and celebrated quite publicly.

Silicon Valley Culture

There are many references to Silicon Valley culture in the popular press. It's even been satirized in a popular TV series called *Silicon Valley*. The founders of many Valley startup companies are justifiably proud of the work cultures they've established within their firms and frequently cite culture as a key contributor to their financial success. So, what exactly is Silicon Valley culture? What's unique about the workplace values, behaviors, and attitudes of Valley employees that differentiates their work cultures from those elsewhere?

First, it's important to recognize that positive and productive work cultures can exist anywhere in any company. Silicon Valley doesn't have an exclusive franchise on healthy work cultures. Second, there's no universally prescriptive formula for defining a good work culture. Good cultures typically possess many common traits but they can also differ significantly from one company to the next. Third, some Valley companies have difficult work cultures that dismay many of their employees or a mix of good and bad traits that inspire some employees and frustrate others. Silicon Valley is certainly not a worker's utopia in which all employees are properly challenged, appropriately rewarded, and deeply satisfied with their jobs.

Valley companies have some common structural characteristics that underpin and help define their work cultures. The majority of companies are relatively small (fewer than 5,000 FTEs), rapidly growing (staff and revenues growing at 30% per year or greater), and relatively youthful (75% of the FTEs are less than 45 years of age). The founders of most companies are still in place and are actively engaged in the strategic management of product plans, recruiting practices, and sales activities. Some have become media celebrities and most are celebrities within their own companies. Many firms employ emerging technologies to develop new types of commercial products and services. They can develop an almost messianic zeal in convincing prospective customers that their products are not simply valuable or desirable but essential to a customer's future prosperity and well-being. New employees sometimes feel that they've joined a cause or crusade instead of a company. These structural characteristics – when considered collectively – set Valley firms apart from larger, long-established companies located elsewhere.

Although I would never claim to be a trained anthropologist, I have observed and experienced a variety of Valley work cultures. I've also had the opportunity to discuss workplace behaviors with other Valley leaders. A fairly common set of cultural traits can be defined on the basis of these experiences and conversations. They are listed below. Individual Valley firms don't necessarily exhibit all of these traits but many or most of these characteristics are present in companies of widely varying size and maturity. Also, as noted above, these traits are not uniquely present within Silicon Valley firms, they're simply more ubiquitous here.

Personal Integrity and Professionalism

Personal integrity and professional conduct are table stakes in any healthy work culture. Leaders have to exhibit integrity in the way they treat others and go about making decisions. Work needs to be assigned, evaluated, and rewarded equitably, following rules and conventions that are understood and accepted by all staff members. Although there have been well publicized exceptions, Valley workplaces do not generally function as clubs, associations, or closed societies with idiosyncratic rituals or enigmatic rules of conduct.

Anxious to Learn

Most IT professionals are well educated but as their careers mature their formal education is superseded by on-the-job experience. Valley professionals have frequently spent more time in the classroom than in the workplace and are anxious to put their formal educations to practical use. Many exhibit a tangible thirst for on-the-job experiences that give them an opportunity to solve real-world problems. They realize that their formal education will only take them so far in realizing their career ambitions. They proactively seek challenging assignments as a means of expanding their knowledge and advancing their careers.

Willingness to Take Risks

Valley workers are generally quite comfortable taking risks. Most work for startup companies trying to establish a commercial toehold for their product or service. Their company's primary objective is to demonstrate sufficient progress to justify the next round of venture capital investment or to achieve financial profitability, so their employees are no strangers to risk. It's equally important to note that there's a significant absence of tradition in many Valley firms, specifically traditions with regard to the way in which work is performed. If a new recruit from another company has a better idea about how to conduct field marketing events, qualify sales prospects, or automate code testing, their new idea will likely be adopted if it can be proven to be better. The classic refrain "that's not how we do things around here" is heard much less frequently in Valley firms than in older, more mature enterprises.

Commitment to Diversity

The diversity issues facing Valley firms in general and technology companies in particular have been well documented in the public press. Senior managers in most Valley firms recognize this problem, are willing to talk openly about it with employees, and are taking concrete steps to expand the diversity of their teams. Ironically, the early success of many companies is partly due to a lack of diversity. Founding teams frequently hire individuals that replicate their beliefs, attitudes, and experiences. This can be useful (maybe essential) in creating a singular focus on constructing an initial product, proving its commercial viability, and determining the size of its prospective market. However, most founding teams would readily agree that self-replication is not a formula for long-term success. They realize that a diverse workforce is essential to scale their company operations in a rapidly expanding market.

Acceptance of Newcomers

Everyone was a newcomer once but for many Valley employees it wasn't that long ago! The trials and tribulations they experienced in acclimating to their new workplace are still fresh in their minds and many go out of their way to provide new recruits with an easier and friendlier onboarding experience. More importantly, newcomers are readily accepted because there's simply so much work to do. If revenues are growing at a rate of 30% or greater, field offices are being opened or expanded every quarter, new products are being introduced semiannually, and smaller firms with interesting technology are being acquired once or twice a year, the existing workforce can be literally overwhelmed by the amount of work that needs to be accomplished to satisfy the demands of the business. New hires are readily accepted in such environments, sometimes more out of desperation than professional courtesy.

Alignment around Winning

Many Valley employees develop a visceral connection with the success of their companies. They truly feel that they're members of a winning team and that they've played a personal role in establishing their company's winning record. Business success can be emotionally validated on a personal level in many ways. It's validated by the addition of new customers and especially by adding a customer with widespread brand recognition. It's validated by external industry awards, the movement of the annual customer meeting to progressively larger and trendier venues, and successive rounds of venture capital funding. Lastly, it's validated by the steady increase in new hires working near your desk until you reach a point where you're moving to a new, expanded office space. These visceral connections to company success are much harder to achieve in larger companies distributed over multiple operating locations that are growing much more slowly.

Personal Connection with Senior Executives

Since Valley firms are relatively small, senior executives tend to be highly visible and directly engaged in day-to-day activities. The founders of small firms commonly conduct weekly all-hands meetings to report recent sales successes, provide product development updates, and enlist the aid of existing employees in filling critical job vacancies. It's not uncommon for these all-hands meetings to persist as a company grows from 50 to 5,000 employees (although the content is likely to change with size). The visibility and communication practices of Valley executives enable many employees to feel that they have a personal relationship with the leaders of their company. While they may not be friends, they've had an opportunity to directly observe the attitudes, behaviors, and beliefs of their leaders. They rely upon these observations to establish an emotional kinship with the leadership team and a sense of trust in executive decisions.

Commitment to Civic Responsibility

In 2001 Salesforce.com established the 1% Pledge model for ingraining philanthropy into the business models of small entrepreneurial companies. The 1% Pledge challenges firms to provide 1% of their equity, profits, revenues, and employee time to civic causes and nonprofit organizations. Many Valley firms have adopted the Pledge and have hired dedicated staff members to ensure its execution within their firms. Many employees participate in the civic programs sponsored by their companies and all take pride in the civic responsibility displayed by their firms. Some companies devote a day of New Employee Orientation to participation in a charitable program that supports a local civic cause.

There's a strong symbiotic relationship between culture and teamwork even though it's difficult to establish explicit cause-and-effect relationships between these two phenomena. Constructive, high-performance teamwork thrives in a healthy work culture. Conversely, productive, mutually supportive teams reinforce and extend the positive attributes of their workplace cultures. Culture and teamwork form a virtuous circle in which a desirable behavior or occurrence in one dimension produces a positive behavior or occurrence in the other.

IT leaders seeking to instill elements of Silicon Valley's entrepreneurial culture within their teams should pursue every opportunity to introduce or nurture the traits listed above. Properly leveraged, these traits are likely to improve the productivity and business impact of their organizations. They're also likely to provide a structural advantage in the future war for talent by making their workplace more appealing to the next generation of IT professionals.

One can legitimately question whether Silicon Valley culture can be sustained in Fortune 500 or Global 2,000 enterprises. Peter Drucker, the famous management consultant, once said: "Culture eats strategy for breakfast." It's probably equally true that growth eats culture for lunch. Valley employees frequently criticize the cultural changes occurring within their firms as staffing levels rise from 50 to 500 or from 1,000 to 3,000. IBM, Cisco, and Oracle were once small, scrappy technology companies with a dream. They were all successful and to varying degrees their success overwhelmed and consumed their initial entrepreneurial cultures. Facebook, Google, and Netflix are all faced with similar problems at the present time. Many Valley leaders would tell you that it's a nice problem to have!

PART II

Process

"It's more fun to be a pirate than to join the Navy."

Steve Jobs

There are too many pirates in IT. Pirates disregard the rules. They thrive on chaos and sometimes incite it. They operate by the seat of their pants and rarely make long-term plans. They welcome mistakes by others as opportunities to advance their own interests. Their on-the-job work behaviors are based on the belief that "it's every man for himself" and they continuously compete for personal fame and recognition. There are too many pirates in IT.

If people are the ultimate source of competitive advantage within an IT organization, process is the means of harnessing their collective talents and transforming that theoretical advantage into practical business outcomes. Examples abound. Successful financial institutions employ rigorous risk management and security processes. Successful retail chains rely on meticulous supply chain management practices. But for some inexplicable reason, IT leaders and practitioners are generally loath to devote time and energy to process management. If conversations regarding people management elicit blank stares from IT leaders, process discussions almost invariably make them cringe. Jobs was probably right – it probably is more fun to be a pirate!

Few IT leaders question the value of process improvement. They simply don't want to become personally involved in managing it themselves. It's far easier to hire process consultants who can instruct their teams on the best practices being used elsewhere in the industry. Many of these same leaders would argue vehemently for budget increases to hire additional staff members. Ironically, they don't understand that the productivity of their existing staff is being compromised every day by chronic deficiencies in process definition and discipline.

Effective processes – clearly defined and consistently enforced – are force multipliers, allowing small teams of people to maximize their productivity by eliminating unnecessary work, avoiding unnecessary crises, and minimizing unnecessary rework. The lack of process discipline traps an organization in an endless stream of self-imposed quality control activities that waste time and delay business outcomes. The absence of clearly defined processes not only undermines staff productivity, it inevitably undermines staff morale as well, as individuals come to realize that they're spending an inordinate amount of time performing haphazard tasks for which they're likely overskilled and unlikely to be rewarded.

Leaders frequently manufacture excuses for acknowledged process deficiencies. Leaders may say that their teams aren't large enough, their internal practices aren't complex enough, or their business partners aren't mature enough to benefit from the introduction of formal processes. This is a blatant exercise in self-delusion. Every IT organization has to support the productivity tools and devices issued to company employees. Every organization has to maintain business applications, manage projects, deal with vendors, secure sensitive data, comply with regulations, etc. There are processes for all of these activities in every IT organization. They just vary in terms of their scope and sophistication. In the absence of disciplined process management, it's likely that multiple processes have been developed in a piecemeal fashion to address different aspects of each activity.

Process management requires leaders to confront the Goldilocks Dilemma head-on. Over-engineered processes never take root within an organization. Some or all of the tasks, activities, communication practices, and documentation requirements prescribed by over-engineered processes are routinely ignored. Under-engineered processes produce marginal benefits precisely because they have significant gaps in their scope or sophistication. Effective management is a continuous balancing act between the scope and sophistication of a given process and the quality of the business outcomes it produces. Leaders need to define and enforce processes that are minimally sufficient to balance organizational effort and operational benefits at any particular point in time.

Minimum Viable Processes

Eric Ries' seminal book – *The Lean Startup* – popularized the concept of a minimum viable product. A minimum viable product is a product whose capabilities are minimally sufficient to test and validate its potential market value. A minimum viable product possesses sufficient business value to justify its continued use and further development. *The Lean Startup* has near-Biblical status in Silicon Valley. Angel and seed investors are constantly encouraging entrepreneurs to validate their business visions by focusing their initial development efforts on the creation of minimum viable products.

The minimum viable product concept is readily adaptable to the Goldilocks Process Dilemma. IT leaders should strive to construct processes that deliver the most business value with the lowest administrative overhead. A minimum viable process is neither over-engineered nor under-engineered. Its scope and sophistication are aligned with an organization's current needs and maturity. Minimum viable processes need to be continuously refined to remain useful. A rapidly growing company will outgrow the application maintenance and end user support practices that were perfectly sufficient a year ago. On the other hand, a company that is reducing overseas operations or eliminating product lines may discover that existing processes are overly complex and unnecessarily burdensome to support its current needs.

A concerted effort to deploy minimum viable processes has important implications. Too many process crusades have been launched and led by well intentioned perfectionists. Perfectionists try to anticipate solutions for every possible edge case scenario. They attempt to emulate the maturity of organizations that have implemented similar processes at larger scales in much more complex environments. They routinely underestimate the training and change management issues involved in introducing new ways of doing business. These chronic compulsions to over-engineer processes can be overcome with a dedicated adherence to minimum viable process principles.

Process crusades are all too often pushed on an organization rather than being pulled into existence by the organization's true needs. Theorists and methodologists construct elaborate frameworks based upon industry best practices. Even when they try to tailor their frameworks to meet an organization's needs, they frequently devote more effort to addressing inconsequential business risks or marginal productivity improvements than is actually warranted. Minimum viable process development is pulled into existence by the most obvious and most significant risk and productivity issues facing the organization. By definition, minimum viable processes are designed to be 80/20 solutions to the organization's problems, sacrificing 20% of the potential benefits to ensure consistent attainment of an 80% gain with the lowest amount of administrative overhead.

A dedicated approach to minimum process development can motivate an organization to embrace continuous process improvement because team members experience the benefits of just-in-time process refinements first-hand. Team members will more readily accept and may even volunteer enhancement suggestions if changes are deferred until their benefits become intuitively obvious to the majority of process participants and stakeholders.

Silicon Valley IT shops employ minimum viable processes out of sheer necessity. Every IT organization complains about insufficient headcount but the staffing limitations imposed on Valley IT groups can be particularly acute. Staffing for product development and sales and marketing functions receive the highest priority in startup firms. Staffing for support functions such as IT are a second priority at best. Valley IT shops routinely establish processes that provide 80/20 solutions to recurring needs. Issues that can't be addressed by such processes are resolved through manual one-off procedures or deferred indefinitely.

Continuous process improvement is a survival skill in rapidly growing firms. It's not uncommon for successful startups to increase revenues by 30% or more on a year-over-year basis. Obviously, the scale and complexity of their internal operations need to expand as well to support such dramatic revenue growth. Startup companies routinely outgrow their existing processes over a period of 12 to 18 months.

Implementing continuous improvement practices within larger, well established enterprises can be equally challenging but in a different way. Large enterprises develop progressively more complex processes to address the growing complexity of their day-to-day operations. Through sheer force of habit, they apply these processes to a wide range of situations and use cases. Some situations benefit from the rigor and sophistication of existing processes while others – sometimes many others – do not. Simply put, the benefits of overly complex processes can be vastly outweighed by their administrative overhead.

Large enterprises usually address the complexity of over-engineered processes in one of two ways. They may create renegade IT groups that are given the freedom to develop customized processes that address their unique needs. For example, renegade groups may be formed to construct a new class of mobile applications, develop an initial set of blockchain-based services, or support their company's eCommerce platform. The development, release, and maintenance practices of such groups may differ significantly from similar processes performed elsewhere within IT. Alternatively, large companies may intentionally simplify or replace existing processes that have become too burdensome over time. It's not uncommon for project approval and management procedures to become more rigorous as the cost and complexity of major company initiatives increase. Many IT shops in larger firms develop "fast track" approval and governance procedures for smaller projects to sidestep the complexity and unnecessary sophistication of their standard management practices.

All too often continuous process improvement is perceived to be an exercise in increasing the scope and rigor of existing procedures. In large enterprises, continuous improvement may have exactly the opposite objective. It may be primarily focused on simplifying and streamlining existing procedures to achieve greater business agility by uncovering the minimum viable process that lies beneath current practices.

In either case described above – small company or large company – the goal is the same: to capture the greatest business benefits while imposing the fewest administrative burdens by either extending or pruning existing procedures.

The Beverly Hillbillies

The Beverly Hillbillies was a popular TV show in the 1960s, later reprised as a movie in the 1990s. It was a situation comedy about a family of Ozark hillbillies who became millionaires overnight after oil was discovered on their property. They moved to Los Angeles and struggled to reconcile their Ozark lifestyle with the cultural practices of their new Beverly Hills neighbors.

In one of my earliest CIO roles, I succeeded an individual who had extensive experience in software engineering. My predecessor had led a large software development team at a Fortune 200 company before becoming CIO at this particular firm whose entire IT team consisted of 500 individuals. Based upon his prior experience he attempted to upgrade the practices that IT employed to develop and maintain our company's business applications. The practices he implemented were far too sophisticated for the work that was actually being performed. They also challenged the technical skills and prior industry experience of our staff members. Consequently, many steps of our formal development and project management processes were routinely ignored. Many of the metrics, documents, and communication practices prescribed by our processes were never produced.

Our COO provided the most apt characterization of this situation. In describing our application development and project management practices, she once said: "Your predecessor created a table setting that would be appropriate for a state dinner with the Queen of England. He never realized that we were just a bunch of Beverly Hillbillies who barely knew the difference between a fork and a spoon!". This metaphor is readily applicable to the over-engineered processes that exist in many IT organizations.

Process Proliferation

In one of my prior companies an offshore team of company employees was established in India to augment the capabilities of the North American IT organization. The time difference between India and North America made it much easier for us to support international business operations on a 24/7 basis. I received a briefing during one of my early visits to India on the ticketing system and performance metrics being used to manage work requests from North America. The system the Indians had devised was effective but differed in several significant ways from the procedures used in North America. When I inquired as to why the Indian managers had decided to develop their own forms and tracking procedures, they told me that they asked several of their North American counterparts about the process being employed in North America. When they discovered that different application and infrastructure teams in North America had developed their own request management processes, they concluded that we had no standardized system and so they simply went off and devised another one on their own. At last count we had five processes for managing requests within different groups. Each had its own set of performance metrics and they employed three different ticketing tools. There's a fine line between letting individual groups develop customized processes that address their unique needs and giving individual teams the freedom to satisfy their personal whims and preferences. I suspect we crossed that line when we unknowingly allowed the Indian extensions of our North American groups to go off and develop their own request management system.

The Process Management Operating Model

There are many well established frameworks describing the critical operational processes that occur within every IT organization. The Information Technology Infrastructure Library (ITIL) service management framework was initially published in the early 1990s and has gone through several subsequent iterations. The most popular revision of this framework is Version 3 which was published in 2007. The Control Objectives for Information and Related Technologies (COBIT) framework for IT operations has matured over roughly the same period of time. The most popular version of this framework is COBIT 5 which was published in 2012. The most up-to-date version of the Project Management Book of Knowledge (PMBOK) was published by the Project Management Institute in 2017. The International Organization for Standardization (ISO) and the U.S. National Institute of Standards and Technology (NIST) maintain widely used frameworks for security management. There are many additional examples.

Frameworks serve as useful points of reference in identifying the fundamental process competencies of any IT organization. However, they are not compilations of prescriptive work procedures. Every organization must define internal processes that address their unique needs and are best suited to their company's size, sophistication, and operational complexity. Frameworks are not cookbooks containing a collection of process recipes. They are books *about* cooking and must be adapted to the needs and abilities of individual organizations.

The ITIL framework referenced above is perhaps the most comprehensive and most widely implemented collection of organizational processes employed within the IT industry. Although it has gone through several revisions, it was initially established in an era in which cloud computing did not exist, DevOps was unknown, work was performed at stationary desktops, and current security and regulatory concerns were unimaginable. Although these subsequent changes in the technology we use and the way we work don't invalidate the ITIL framework, they force us to refashion and reprioritize the ways in which ITIL processes are implemented in the future.

It's not the purpose of this book to review existing frameworks or pass judgment on their respective strengths and weaknesses. Instead, the following discussion will selectively focus on nine processes that need careful nurturing or potential redesign to operate successfully during the next decade. Some of these processes represent adaptations of well known ITIL and COBIT processes. Others are squarely focused on gaps in the traditional frameworks.

Service Management

The term "service management" can be used in many different contexts but for the purpose of this discussion it refers to IT's ability to fulfill end user requests, resolve end user incidents, diagnose systemic problems giving rise to such incidents, and put changes into production environments (software and hardware) that remediate problems. In brief, it refers to the classic ITIL processes of Request, Incident, Problem, Change, and Release Management.

Effective service management is a foundational competency in any IT organization. Business executives may gauge the organization's success on its ability to complete major projects on time and on budget but the remaining 90% of the company's employees will measure IT's success on its ability to provide the resources they need to perform their jobs. IT may play a leadership role in implementing new technical capabilities and enabling strategic business initiatives but at its core it is and always will be a service organization. *Effective service management is a necessary but not sufficient condition for establishing the business credibility of every IT organization.*

There are three critical dimensions to every service interaction that an end user has with IT: technical competence, personal empathy, and time to resolution. Each dimension is essential to a successful interaction. First and foremost, end users seek to determine if the IT staff has the knowledge and skills required to resolve their issue or satisfy their request. Secondly, they subliminally assess the level of attention, personal rapport, and sense of urgency displayed by the IT staff members addressing their concern.

Finally, every end user will subjectively determine whether their issue or request was resolved in a reasonable period of time employing their own personal definition of "reasonable."

The technical challenges involved in resolving IT issues frequently obscure the fact that service delivery is an intensely personal experience. There's an old saying that "there's no such thing as minor surgery when it's you and the knife." The same is true regarding IT. There's no such thing as a minor IT problem if that issue is keeping you from doing your job. Too many IT organizations focus almost exclusively on the technical dimension of their interactions with end users and fail to actively manage the emotional dimension of such interactions.

Successful end user interactions have to win on all three counts. The successful resolution of a technical problem by a surly or arrogant Service Desk agent is not a win. However, an empathetic interaction with an agent who lacks the technical competence to satisfy a request is also a loss. The successful resolution of an issue by a highly empathetic agent after a week-long exchange of emails and text messages would also likely qualify as a loss, except in the case of the most complex incidents or requests. In short, there's far more involved in winning the hearts and minds of IT's customers than simply closing Service Desk tickets in a timely fashion.

The vast majority of IT–employee interactions are funneled through the Service Desk. Successful Desk operations involve tradeoffs between technical competence, response times, and cost. The most successful interactions typically occur at walkup Desks distributed throughout an office campus that are staffed by highly skilled agents. However, this is the costliest solution. Alternatively, support for the resolution of routine issues can readily be outsourced, usually at a lower cost.

There's a growing reliance on technology to satisfy end user support needs. Knowledge management tools and chatbots can provide self-service solutions to common problems encountered by many employees. Online service catalogs contain automated routines for provisioning applications, resetting passwords, and ordering monitors. Vending machines dispense commonly used equipment. Employees obviously value timeliness and convenience in resolving their issues but all too often they need to explain the unique features of their problem or request to a human being.

Mechanisms for deflecting issues from the Desk can be useful when properly focused. However, IT needs to avoid becoming overly reliant on these deflection mechanisms. In many instances they're a poor substitute for direct, personalized interactions and may fail because: they're unable to capture a complete depiction of an end user's issue; they don't possess the information needed to fully resolve the issue; they lack an empathetic understanding of the importance of the issue from the end user's perspective; or their prescriptive guidance confuses the end user, resulting in protracted resolution times.

Desk agents need continuous training to cope with the wide variety and constantly changing nature of the technical issues they encounter. Since successful service delivery has a significant emotional dimension, they also need training in the soft skills involved in managing service interactions. By way of example, Starbucks has a structured interaction methodology for resolving customer problems that it refers to as LATTE:

- **L**isten
- **A**cknowledge the problem
- **T**ake problem-solving action
- **T**hank them
- **E**xplain what you've done

While this particular methodology may not be directly applicable to Service Desk operations, it underscores the importance of establishing a systematic approach to managing individual interactions with the Desk's customers, particularly in contentious situations.

Communication plays a critical role in successful service delivery. Too many IT organizations are referred to as "black holes" by their customers. Incident reports and service requests are submitted with little if any indication as to when they'll be addressed or resolved. Service disruptions triggered by network problems, SaaS outages, or emergency system maintenance frequently go unannounced, prompting a flood of calls and tickets to the Desk. Even after such disruptions are resolved, there's no follow-up communication from IT explaining what went wrong and what steps are being taken to avoid a reoccurrence of the same problem. There are many ways of improving IT's interactions with its customers simply by communicating more immediately, more frequently, and more substantively. A wide variety of collaboration tools are readily available that can focus the delivery of relevant service information to specific individuals, teams, or communities. In the next decade, IT organizations should review their end user communication practices on a regular basis to ensure that existing practices are still appropriate and are being consistently enforced.

Seven Deadly Sources of IT Dissatisfaction

The credibility of an IT organization can be undermined in many different ways. Major disasters such as the loss of the primary network switch in the corporate data center, the failure of retail point-of-sale terminals during the holiday shopping season, or the corruption of data within a company's data warehouse leave lasting impressions. Chronic failures in delivering major projects on time and on budget can create doubts about IT's managerial and technical expertise. Equally damaging but far more insidious are the day-to-day hassles that employees routinely experience in their personal use of IT resources. It *is* the little things that frequently drive employees crazy!

In no prioritized order, the seven deadly sources of chronic employee dissatisfaction with IT are as follows.

Passwords. At best they are a nuisance. At worst they are a nightmare. Employees constantly complain that passwords are too long and too complex to remember and are not properly synchronized across multiple applications. Furthermore, they complain that they're asked to enter and change passwords far too frequently.

Trouble tickets. IT thinks of Service Desk trouble tickets as a logical system for prioritizing end user support activities. Employees view tickets as a gauntlet that must be endured to obtain a resolution to their problem or request. Ticketing systems are viewed as "black holes" by many employees: once their issue has been submitted as a trouble ticket, they have absolutely no idea if, when, how, or by whom it will be resolved.

Video conferencing. How much time is wasted in large enterprises trying to get conference room video equipment to work properly? How many times do users revert to a phone bridge after wasting the first 10 minutes of a meeting trying to follow the video setup procedures supplied by the IT team? The answers: too much and too many!

Distribution lists. In a perfect world, distribution lists are always accurate and easy to construct. Up-to-date information concerning an individual's employment status, job title, cost center assignment, and organizational reporting relationship is readily available from their company's human resource system. In a perfect world, accurate distribution lists can be created dynamically at the touch of a button. Unfortunately, very few people live in this perfect world. Most distribution lists must be manually reviewed for accuracy before they're actually used.

Sluggish response times. Employees who have been conditioned to expect split second response times in their use of consumer websites such as Amazon, ESPN, or CNN have a low tolerance for slow application response times at work. Sluggish response sometimes occurs on a predictable basis as Wi-Fi access points or back office transaction systems experience peak loads at specific times of every day, every week, or every quarter.

Blocked websites. Employees experience unbounded freedom outside the workplace in using the Internet to obtain information, make purchases, socialize with friends, and amuse themselves. Subliminally, they feel they're entitled to the same freedom at work. Many are shocked and personally offended when their right of access to an external website is blocked by IT.

Printers. Perhaps the most chronic frustration of all time, paper jams, empty ink cartridges, and printers waiting for parts are inevitably encountered at the most inconvenient moments, typically before major presentations or document filing deadlines. (Ironically, printer-related frustrations are waning rapidly – not through the use of new printer technology but by the digital transformation of the workplace itself into a completely paperless environment. Printer-related problems have been largely resolved by the wholesale rejection of paper as an efficient medium for business communication and collaboration.)

Fortunately, there have been significant technical advances in many of these areas. Automated password reset procedures, walk-up help desks, user friendly videoconferencing systems, and intelligent printers that issue maintenance requests in advance of failure address many of these recurring issues. Some organizations have made significant progress in resolving several of the problems discussed above, but few – if any – can claim to have resolved them all.

Effective service management processes are not only designed to respond to employee support issues, they're also designed to minimize or avoid such issues altogether. Recurring incidents may be symptoms of more systemic problems that should be resolved through a formal problem management process. Other incidents may result from the implementation of new capabilities that have not been properly tested or vetted across the IT organization itself. Formal configuration and change management processes can correct many of these problems. Periodic degradation in the reliability or responsiveness of key systems can be mitigated through more formal availability and capacity management processes. All of the classical elements of the broader ITIL service management framework will remain relevant in the next decade, even in a hybrid world of cloud-based and on-premise operations.

Over the past 5 years it's been fashionable to debate whether DevOps practices have superseded and replaced conventional service management processes. DevOps teams perform many of the functions described in the ITIL framework. They configure infrastructure, monitor application availability, respond to user requests and incidents, and make continuous changes to production systems. DevOps aficionados have argued that ITIL principles are simply embedded in DevOps practices and don't need to be managed through a separate, structured framework. That's one of the recurring problems with frameworks – they can fuel semantical debates that are entertaining but irrelevant. ITIL principles will continue to underpin the operations of every IT organization in the next decade, regardless of how they are adapted, codified, and enforced.

Application Development and Maintenance (ADM)

Classic requirements for ADM support still exist in a cloud-based world in which business applications are purchased as subscription services and proprietary applications are hosted on a cloud vendor's infrastructure. The biggest – perhaps revolutionary – change in ADM during the next decade will be the wholesale adaptation of Agile principles to maintain and enhance the unique application portfolios employed by individual companies. Waterfall principles may still play a role in implementing large IT initiatives through a series of smaller, phased projects but the underlying work performed within each project will be managed through the use of Agile processes.

ADM remains relevant in a SaaS-dominated environment because businesses continue to customize SaaS applications in much the same way they customized on-premise applications in the past. ADM teams add data fields to standard objects, construct custom objects, configure feature flags, expose APIs to other services, build software packages, and extract data from other applications via APIs. As customizations expand in scope and complexity, greater degrees of quality assurance (testing) are required before new enhancements can be implemented in production.

Changes to major business systems have historically been implemented in multi-month cycles. Cycle times have typically varied from 2 to 4 months. Quarterly (3-month) cycles were particularly popular for enhancing enterprise resource planning (ERP) systems. These cycles were typically completed in the middle of a fiscal quarter to ensure that business teams had sufficient time to test enhancements before all changes to critical systems were suspended prior to quarter close (commonly referred to as the quarterly blackout period). The timing of changes to other systems was frequently synchronized with the timing of ERP changes due to their dependencies on the ERP system itself or as a matter of convenience.

Agile principles have had a disruptive impact on these historical practices. Agile was initially proposed as a software development framework but its underlying concepts have been extended and applied to many other types of IT activities, including ADM. Philosophically, Agile is designed to promote sustained collaboration among IT team members and their business partners, deliver predictable results in an incremental fashion, and establish a continuous improvement mentality among team members. In practice, it's commonly implemented as a cyclical scrum process that delivers results through a series of successive sprints.

ADM sprints are planned and completed over relatively short periods of time, typically on the order of 1 to 4 weeks. Each sprint allocates time to fixing current performance issues within production systems, performing sustaining engineering activities required to maintain system health (i.e. enhancements or upgrades not requested by business partners), and implementing business-requested enhancements. The scrum process may also be used to implement new applications or re-engineer business processes, in which case the individual sprints contain a mixture of the design, development, testing, and change management activities required to put new systems or business capabilities into production.

Although Agile is commonly employed to manage routine ADM activities, it should be used with caution. There are many situations in which it's not appropriate or needs to be augmented with other management practices. For example:

- Agile is not necessarily an efficient means of managing ad hoc administrative requests such as modifying application access privileges, generating one-off reports, making minor configuration changes to an application or applying minor corrections to an application database. Although administrative requests may appear to be relatively inconsequential and easy to defer, they can be critically important to a requestor. Failure to fulfill such requests in a timely fashion may impede the requestor's ability to perform their job. The time required to catalog, prioritize, and schedule such requests within successive scrum cycles may be totally disproportionate to the time required to simply perform the requested task. Administrative task requests can be managed in a more efficient manner through the use of a ticketing system that prioritizes requests on the basis of their business urgency and age.

- Business partners may not be interested in implementing a series of incremental application enhancements because the individual enhancements have marginal business value or the change management issues involved in their immediate use are too burdensome. In many cases business partners prefer to bundle multiple enhancements and implement them collectively at a single point in time. Enhancements may still be constructed through a series of sprints but business leaders may prefer to consume such enhancements on a periodic Waterfall basis.

- Aging systems may generate a continuous stream of production support issues that overwhelm ADM scrum cycles. Time normally reserved for enhancements and sustaining engineering may be routinely displaced by pressing support issues. Under these circumstances a Kanban scheduling system in which support issues and other activities are continuously reprioritized may be a more appropriate management process.

- Some business applications such as Salesforce.com are structured as platforms that can be readily expanded through the addition of application packages. These packages are built upon a platform's Application Programming Interfaces (APIs). Other applications are primarily customized through configuration changes which require little, if any, code development or interface testing. As in the case of administrative request management, Agile processes may impose a procedural burden on the enhancement of configuration-intensive applications that adds little value and actually slows the delivery of new business capabilities.
- Agile processes may be too narrowly focused on code development and test practices. They may fail to address all the operational and business issues that are involved in employing new software capabilities. New enhancements may impact application performance, raise security issues, require compliance with regulatory controls, entail unique forms of monitoring and operational support, or necessitate the re-engineering of existing business processes. These non-software issues need to be resolved before a software enhancement can deliver true business value.

Agile has emerged as the predominant framework for managing ADM. Its benefits clearly outweigh any of the potential limitations referenced above. Agile can significantly improve a team's productivity by putting a sharp focus on short-term commitments and minimizing the constant reprioritization of work activities. Execution risk is reduced simply because team members are far more capable of determining what they can realistically accomplish during the next 2 weeks than the next 2 months. Productivity improves because team members make a collective commitment to complete a specific set of tasks during each sprint cycle. There's far more peer pressure involved in achieving agreed upon objectives under these circumstances than in simply making individual commitments to a team's manager. Finally, short scrum cycles sustain a higher level of business partner engagement, both in terms of prioritizing the work to be performed and leveraging the results. Agile has proven to be a resilient methodology for performing routine ADM work.

Business partners are mostly indifferent to the nature of the ADM work practices an IT organization chooses to adopt. All they are seeking are predictable (and frequent) business outcomes. Their measure of success is whether IT can consistently meet its commitments and produce results that have business significance. In many cases business partners welcome the adoption of Agile simply because it provides a means of holding their IT counterparts more accountable.

Agile is not a silver bullet. Agile practices may actually make an ADM team appear less responsive to its business partners. In some situations – such as the ones referenced above – Agile may deliver an incomplete solution to the true needs of the business. It can (and should) make it more difficult for business partners to change priorities or introduce new requests during individual sprint cycles. It will test IT's ability to accurately estimate the work required to complete specific tasks and activities. Chronic failures to complete the work assigned to individual sprints will displace work planned for subsequent cycles, aggravating the frustrations of IT's business partners. Agile processes magnify the accountability of IT teams and make failures to perform much more visible, both to the team members and their clients.

Agile is a framework of work principles, not a theology. These principles need to be carefully adapted to address a specific set of ADM challenges. They should not be applied on a wholesale basis to address all ADM needs. Agile may be used to facilitate the incremental discovery of business requirements, expand business involvement in the prioritization of enhancement requests, reduce execution risk, accelerate the delivery of new capabilities, improve team productivity, or reserve dedicated time for sustaining engineering activities. Agile practices developed for individual organizations need to target one or more of these objectives and establish metrics to ensure they are being truly achieved.

Product Engineering Principles in IT?

Can the engineering principles used to develop commercial software products be applied to the development and maintenance of business applications?

Most Valley companies are software factories. Product Managers establish product roadmaps describing the features and functions that will be added to existing products or used to develop wholly new ones. Roadmaps are translated into a set of technical specifications that are implemented by Product Engineering teams, typically using Agile principles and DevOps procedures. Small specialized engineering teams focused on individual products are widely considered to be the most efficient means of performing this work.

At first glance, the ADM practices within most IT organizations are somewhat similar. IT application teams receive feedback from business users regarding the shortcomings of current systems and future operational needs. Individual teams employ a variety of processes to convert this feedback into a prioritized list of business requirements. Requirements are translated into technical specifications that may involve changes to the configuration of existing applications or the development of customized software packages. Small specialized teams with in-depth knowledge of individual SaaS platforms are typically considered to be the most efficient means of performing this work. For example, a Salesforce.com ADM team is usually established not only to configure the Salesforce application itself but also to manage other applications and customized software packages that are built upon Salesforce's APIs.

Although there are some general similarities between commercial product development and business application support, there are also some important differences that limit the wholesale application of product engineering principles to ADM. Key similarities and differences are discussed below.

Customer Interactions

Product Managers base their roadmap plans on information from multiple sources. They try to appease the desires of their largest

customers, surpass the capabilities of key competitors, and leverage new forms of technology to enhance existing products or create
wholly new ones. They distill their roadmaps into a set of functional
requirements that can be implemented by their company's Product
Engineering teams. Progress is typically measured on a quarterly basis.
Roadmap plans may be revised or revisited annually or semi-annually. The Engineering teams themselves have limited interaction with
current customers, competitor products, or their company's senior
product strategists. They effectively take their marching orders from
the Product Managers.

ADM is a far more intimate affair. Application teams receive feedback from a wide variety of end users and from multiple business
functions. Team managers are directly involved in facilitating the
prioritization of user requests and may need to engage senior business leaders in adjudicating conflicting priorities. Business Systems
Analysts work directly with their business counterparts in clarifying
requirements and testing new features and configurations. Application
developers may even have direct interaction with business users during
scrum meetings. In most instances, business users are directly involved
in testing and approving changes to existing applications. The ultimate measure of intimacy between application team members and
their business clients is the fact that they work for the same company
and are frequently located in the same office. ADM team members can
interact with their customers informally in the company parking lot or
cafeteria, and frequently do so.

External Team Dependencies

Commercial DevOps teams are designed to be as self-sufficient as
possible. They manage the infrastructure resources required for code
development and testing. They enforce prescribed coding and testing
procedures. They release new code into existing production systems.

They monitor the performance of production systems and take corrective action as needed.

ADM teams are largely self-sufficient with regard to configuring their applications and constructing custom software packages. They also resolve production support issues directly or enlist the aid of their vendors in fixing production problems. ADM teams typically rely on the expertise of data warehousing, application integration, and information security specialists that reside outside their teams to complete many work assignments. ADM teams have minimal infrastructure responsibilities because their applications are hosted by their vendors in the cloud. However, they may need to consult with their vendors or employ the services of a vendor's professional services team before certain projects can be completed. ADM teams generally have more external dependencies than Product Engineering teams and are less self-sufficient.

Planning Horizons

Product roadmap plans evolve over time in response to changing customer demands, new competitor capabilities, or the emergence of new technologies. Roadmap plans are typically reviewed and updated annually or semi-annually. They may be revisited spontaneously if a competitor introduces a new capability that threatens a company's current or future market position. ADM enhancement plans are typically more fluid and may be revised quarterly or even more frequently in response to changing business priorities. ADM teams need to be more dexterous in juggling the ever-changing and sometimes conflicting demands of their business partners to ensure that they are working on the most business-critical issues and opportunities at any point in time.

In summary, although the nature of the work performed by Product Development and ADM teams is generally similar, the ways in which work is planned and performed are quite different. Both teams develop, test, and release code. Both teams configure systems and resolve production issues. However, the ways in which they interact with their users, employ external expertise, and prioritize their work plans are significantly different.

Debates regarding the applicability of product engineering principles to ADM activities all too frequently degenerate into semantical arguments in which the combatants simply use different terminology to refer to the same type of role, activity, or function. The real value of product engineering principles within ADM is the adoption of Agile scrum practices to plan and manage the work performed by application support teams. The benefits of Agile are discussed elsewhere in this book. They can be realized as easily within ADM teams as they have been by Product Engineering teams in the past.

API Governance

Application Programming Interfaces (APIs) have become far too significant from a business perspective to be relegated to the technical toolboxes of software developers. Business APIs expose critical business data or algorithms to internal systems or external stakeholders. For example, they may provide a customer's current billing address to an internal system or the result of a calculation, such as a personalized mortgage loan rate, to a mobile consumer application.

APIs serve many useful purposes. They provide a powerful communication mechanism for exchanging data and information among multiple applications that support a common business process. In so doing, they avoid the need to duplicate data and business logic in multiple systems and simplify the identification of master sources of critical business information. They can enhance the productivity of software development teams by avoiding duplication of effort and accelerate the time-to-market of new IT-enabled business capabilities. Most importantly, they can provide employees, suppliers, partners, and customers with dynamic access to near-real-time data and information.

At a strategic level, APIs may extend a company's business model, allowing firms providing complementary products or services to leverage data and information that was formerly restricted to a company's internal systems. Conversely, a company may be able to extend or enrich its current product offerings by incorporating data or services furnished by external third-party APIs.

Although APIs are routinely considered to be an effective tool for integrating disparate applications, they can play an even more significant role in achieving deeper *business integration* between a company and its suppliers, managed service providers, go-to-market selling partners, contract manufacturers, distributors, franchisees, affiliates, etc. APIs can also facilitate broader and deeper interactions with a company's customers. Information sourced from external APIs can increase the relevance and intimacy of digital interactions with customers, promoting more frequent, personalized, and rewarding customer experiences. In strategic terms, APIs are ecosystem catalysts that enable companies to expand their operational boundaries and overall business impact by facilitating the sharing of data and information.

An API-based approach to application architecture is a paradigm shift from the construction of monolithic applications that are wholly self-sufficient to the assembly of applications that integrate data and services from both internal and external sources. In layman's terms, it's the difference between weaving a tapestry and stitching a quilt!

Almost every company has an API architectural framework, whether they realize it or not. SaaS applications are largely API-based. Many of their APIs are public and available to a company's application support team. Access to other APIs may be restricted to a SaaS company's partners. Partners commonly leverage non-public APIs to develop their own commercial products and services. So, even if a company doesn't have a formal API strategy, their existing business applications may already be exchanging data and providing services via APIs.

As APIs assume a more critical role from both an IT and business perspective, they need to be managed more intentionally and strategically.

Enterprise APIs that support sensitive business operations or are widely used by multiple applications need to be standardized, documented, versioned, and cataloged. Their use needs to be consistent with operational controls that have been established to comply with SOX, PCI, ISO, HIPAA, FedRAMP, and GDPR regulations. End user authentication procedures and access privileges need to be explicitly enforced at or upstream of the API interface and verifiable via system logs. A comprehensive catalog of enterprise APIs will include both internally developed and externally available APIs.

API utilization needs to be monitored on a routine basis and used to curate a company's API collection. Frequently used APIs may be too complex. Their use may trigger unnecessary computations or multiple database calls that introduce latency and generate nonessential network traffic. Frequently used APIs may be deconstructed into two or more simpler APIs that avoid these unwanted side effects. Conversely, it may be possible to merge the functionality of occasionally used APIs to reduce API proliferation. Some APIs may be retired altogether over time while others may be simplified to minimize redundant functionality.

Responsibility for the management practices outlined above – standardization, documentation, catalog maintenance, operational monitoring, security compliance, and curation – cannot be distributed among multiple teams using different tools, procedures, and repositories. In the operating model of the next decade, dedicated teams will be established to define, standardize, and maintain enterprise business APIs. API reuse will be promoted through thorough documentation and standardization, guaranteed adherence to regulatory controls, and continuous utilization monitoring. These teams will require investments in tools and personnel but will ultimately pay major dividends in expanding the utility of SaaS applications, accelerating the time-to-market of new software-based products and services, deepening business integration with external stakeholders, and enriching customer experience.

A wide variety of API management platforms currently exist. These platforms provide the technical mechanisms for creating and publishing APIs, enforcing API access policies, queuing API calls, and monitoring API utilization. In the absence of enterprise-level governance practices, however, these platforms will only provide tactical operational benefits.

Every company will truly be a technology company in the next decade. Most will achieve competitive advantage through the software capabilities they assemble and integrate. *It's inconceivable that companies seeking to digitally transform their businesses will be able to do so in the absence of explicit and systematic API governance practices. The technical and business synergies that can be realized through deeper, more pervasive integration of a company's software assets are simply too great to be ignored.*

Data Management

Timely, accurate, consistent data has always been important to IT's business partners. As application and infrastructure resources become commodities that are readily available to everyone, the proper management of a company's data resources becomes an increasingly critical source of competitive business advantage. One could easily argue that in the final analysis, the sole purpose of the applications and infrastructure that IT maintains is to deliver the right data to the right business person at the right time.

Data plays two important roles in business operations. It is injected into transactional processes to produce accurate and timely business outcomes. For example, accurate billing information is needed to prepare a monthly customer invoice. It's also analyzed to inform strategic business decisions. For example, a historical analysis of seasonal demand for different types of women's apparel might be used to revise a company's discounting strategy during the next holiday shopping season. In the former instance data is being used for tactical purposes on a transactional basis. In the latter instance it's being used to alter future business plans. In either case, data accuracy, consistency, and timeliness are essential.

The data management challenges in most companies are so complex and have existed for so long that they're routinely ignored or sidestepped by IT leaders. It's not uncommon for IT initiatives to focus inordinate attention on the usability of a new system, its global availability and response times, and its resiliency in the event of an equipment failure. But they rarely become concerned with data quality issues. When new systems require access to legacy data it's usually imported directly, without regard for preexisting gaps, errors, or inconsistencies. Furthermore, the data integrity issues associated with legacy systems are regularly allowed to propagate into companion systems, creating additional management issues and reporting inconsistencies.

Data management issues in Valley companies are aggravated by the proliferation of SaaS applications. Functional teams frequently implement a variety of SaaS tools that contain similar or identical data. These applications routinely exchange data with one another in an ad hoc fashion that isn't controlled or coordinated via some type of master schedule. The lack of consistent data definitions, failure to identify master data types and sources, and ad hoc exchange of data among multiple applications can undermine the integrity of selected business processes, the accuracy of analytical projects, and the utility of standard business reports.

Data management is not a program, or a set of business rules, or a collection of procedures. It is all of the above. Proper stewardship of a company's data resources requires an institutional commitment to data integrity. IT alone can't guarantee the accuracy of data being entered into source systems by different functional groups. Data integrity can become a cultural norm in any company whose executives demand timely access to accurate data to support daily operations and make strategic decisions. If business executives are only marginally interested in using data to improve business efficiency and decision effectiveness, then IT is unlikely to develop meaningful data management processes on its own.

The key challenges involved in effective data management are discussed below. None of these are new or revolutionary. However, they are all compounded by the proliferation of SaaS applications and their associated data stores.

Role of the enterprise data warehouse (DW). During the next decade, operations teams embedded in individual functional areas will continue to administer many aspects of their core SaaS applications. Functional administrators may configure features and workflows within such applications, load data, manage fine-grained access privileges, and publish routine reports. Functional analysts may use application-sourced data to perform a variety of strategic studies as well. In contrast, an enterprise data warehouse aggregates data from multiple applications. It is used to address enterprise business needs that span multiple functional teams. There's almost always an inherent tension between the accuracy and utility of application-sourced reports created by functional teams versus warehouse-sourced reports created by the DW team. Many of these tensions can be traced to differences in data normalization routines, data definitions, and data synchronization practices. There are proper and necessary roles for both forms of reporting. However, the line of demarcation between their respective roles and responsibilities needs to be explicitly defined and consistently enforced. This is particularly challenging within Valley firms that are continually implementing new SaaS applications and continually adding new data sources to their data warehouses.

Master data dictionary. Key data elements of the greatest business significance need to be defined consistently across the enterprise and represented in data models that portray their interrelationships. A master source system for every element that meets this definition also needs to be formally designated. Standard computational practices need to be defined for derived business terms such as "average purchase price" or "lifetime customer value." Master data management (MDM) initiatives have historically floundered because their scope, length, and cost expand beyond the tolerance of their sponsors. MDM initiatives are ideally suited for Agile implementation practices. It's best to start small, focus on elements that are universally considered to be business-critical, and expand the dictionary incrementally over time. It will be easier to enlist participation by others once the benefits of the dictionary become readily apparent.

Data integrity. There's no polite way of saying it: there's a lot of crappy data in most IT systems. Data may be improperly or inconsistently entered into systems by improperly or inconsistently trained users. Significant data gaps may exist because certain information wasn't available to a system's users at the time of data entry. Certain forms of data have a shelf life. They age over time and become progressively less accurate if not validated and updated on a regular basis. Data definitions may have changed over time as well, creating additional inconsistencies. And finally, databases are filled with information that hasn't been used or accessed for years. Formal data retention policies don't exist for many systems and are not always enforced where they do exist. There are a variety of tools and procedures available to clean, correct, and normalize data, including some emerging tools that rely on crowdsourcing techniques to validate data integrity. In reality, data integrity issues are never completely solved but their threat to successful business operations can be reduced to acceptable levels.

Data security. There's a fundamental conflict between IT's mission to provide its business partners with easy access to critical information under a wide variety of circumstances and its responsibility to protect sensitive company information. It's wholly unrealistic and borderline impossible to provide blanket protection for all business-critical information. Investments in security safeguards must be tailored to the business risks associated with the potential theft of different types of information. Regulatory requirements are actually helpful in classifying the sensitivity of data housed in different systems. SOX regulations apply to systems containing confidential financial information. GDPR regulations apply to systems containing Personally Identifiable Information (PII). Most companies are already acutely aware of systems that contain their proprietary intellectual property (IP). An understanding of the sensitivity of the data contained within different systems is required to tailor appropriate security safeguards for their use.

The widespread adoption of SaaS applications creates security benefits and concerns. Many SaaS vendors employ security safeguards that are far more sophisticated than those employed by the majority of their customers. Under these circumstances, company information is actually *better protected* within the SaaS vendor's system than it would be otherwise. On the other hand, data can be easily accessed and transferred among multiple SaaS applications, making it even more difficult to maintain an accurate data classification framework and standardized set of data definitions.

Inaccurate or incomplete classifications and ambiguous definitions will compromise the effectiveness of safeguards that have been put in place in the past.

Self-service. IT will never be able to fund the army of analysts that would be required to satisfy all the data needs and desires of its business partners. Self-service data access is the only solution to this dilemma. Although it appears to be a tractable problem in theory, it's devilishly difficult to achieve in practice because of the widely varying needs, skills, and sophistication of different data consumers. The number of tools available to analyze data, generate routine reports, construct dynamic dashboards, and visualize information has grown enormously over the past 5 years. Individual functional groups are likely to select a unique set of self-service tools that they believe is best suited to their needs. IT's primary job is to ensure that it can deliver clean and accurate data to such tools on demand.

The data management challenges summarized above are all interdependent. They can't be neatly organized into a prioritized checklist of serial initiatives. No organization has the luxury of working on these challenges one at a time and no organization can declare that it has permanently resolved any of them. Data management challenges evolve as a company's business model evolves and its IT infrastructure changes.

Data management is no longer an unavoidable hygienic activity performed within the bowels of the IT organization. It has a critical impact on employee productivity, accurate financial reporting, business agility, and customer satisfaction. Consequently, it should be at the forefront of the process agenda of every IT leadership team. It requires their immediate and sustained attention.

Security Culture

It's fashionable these days for CEOs to proclaim that "every company is a technology company." It's equally true that every company is a security company because the theft of critical business information could readily result in a loss of revenue, customers, or creditworthiness. CEOs are reluctant to proclaim that every company – including theirs – is a security company but in their hearts they know it's true.

It's difficult to determine if cybersecurity should be discussed in a people context, process context, or technology context. All of these factors play a critical role in safeguarding the information assets of a modern enterprise. Security is classically considered to be the end result of investments in specialized expertise, technology tools, and operational procedures. In reality, it is much more than the simple sum of these organizational capabilities. Cybersecurity needs to become an overarching cultural norm that pervades a company's strategic plans, tactical initiatives, and daily business activities.

Cybersecurity is not solely the responsibility of the IT organization. It needs to become an integral part of the operating DNA of the enterprise. In the same way that employees should continually challenge the effectiveness of current work practices and future business plans, they should be challenging the effectiveness of existing security safeguards. They should be acutely aware of their company's cyber risks and fully understand the potential business implications of failing to address their cyber vulnerabilities. Security awareness should permeate the conduct of everyday activities and routine decision-making.

While IT cannot establish an enterprise-wide security culture on its own, it should serve as an example for the rest of the corporation. Unfortunately, this is rarely the case. There are too many IT shops in which security responsibilities have been delegated to a small team of security professionals and are largely ignored by other staff members. Many IT groups outside the security team routinely dismiss, disregard, or debate instructions to insert more rigorous safeguards into their current operational procedures or existing technology stacks. Furthermore, it's not uncommon for individual staff members to express dismay or indifference when asked to assist in resolving a security-related audit issue or responding to a security incident.

Although IT can't create an enterprise-wide security culture on its own, it's highly unlikely (borderline impossible) to create a company security culture if one doesn't already exist within the IT organization. IT manages too many information assets and controls too many access channels to abdicate its responsibility to foster enterprise-wide security awareness. In the next decade, successful IT organizations will develop a broader, deeper, and more pervasive understanding of their cyber vulnerabilities. Team members throughout the organization will proactively identify security issues associated with routine operations and the implementation of new capabilities. They will also take personal responsibility for aggressively enforcing existing safeguards.

Many readers of this book may argue that it's impossible to truly establish an enterprise-wide security culture. They may claim that such a vision is a self-serving pipe dream promoted by security zealots. There's an abundance of empirical evidence that categorically refutes this perception. Pharmaceutical companies developing proprietary drug formulas have security cultures because their financial success depends upon the protection of their intellectual property. Legal firms managing high-stakes litigation or highly competitive M&A activities have security cultures to ensure their professional reputations and future business prospects aren't compromised. Personal wealth management firms for some of the world's most successful businesspeople and celebrities have security cultures. In practice, it's actually easy to establish such cultures once risks are clearly understood and the business consequences of public breaches are patently obvious.

The formula for establishing a comprehensive *security program* is well known and consists of the following building blocks:

- *Understand your cyber risk landscape.* What collection of hackers, criminals, and nation states would likely be interested in your information assets or gain in some fashion by breaching your security defenses? How has this collection of threat actors changed over time as your business model has evolved? The risk landscape facing your enterprise should be reviewed and revised on a regular basis, perhaps every 9, 12, or 15 months. It should not be treated as a one-time exercise to be revisited every 3–5 years.
- *Understand your cyber vulnerabilities.* This is a three-part exercise that starts with an understanding of the databases, systems, and business

practices that create, store, or consume your company's most sensitive information. Once these targets have been identified and ranked in terms of their business significance you need to identify the various ways in which they can be accessed. Finally, the business consequences of having any of these targets breached should be quantified to the maximum extent possible.

- *Establish policies to address your vulnerabilities.* All too often, companies establish generic security policies that are only loosely mapped to their true technical and business vulnerabilities. This results in extraneous effort being devoted to protecting lower vulnerability targets at the expense of the heightened focus that should be applied to the highest vulnerability targets. Security policies should be customized to address the unique features of a company's business model and operational practices.

- *Establish the team, tools, and operational practices to enforce and audit your policies.* This is the work that IT classically does well but its efforts may be ineffective if the preceding building blocks have not been carefully constructed.

A comprehensive security program is a necessary but insufficient condition for the creation of a successful *security culture.* The formula for a successful culture requires all of the above elements plus the vigilant intellectual and emotional engagement of every IT staff member – from the most senior Enterprise Architect to the most junior System Administrator. The tried and true practices used to plan major IT initiatives can't be directly applied to the creation of a security culture. A true cultural revolution isn't planned, budgeted, and scheduled. It's a crusade dedicated to winning the minds and hearts of individual team members. While there's no one-size-fits-all formula for success, here are some triggers that can help start the revolution or propel it forward:

- *Teach.* Establish formal and informal mechanisms for providing team members with more insight into the threat actors that would profit from breaching your defenses. Share stories about other companies whose defenses were breached and the business consequences they experienced. A major breach will disrupt your team's current tactical initiatives and strategic plans. Team members need to understand that their existing projects and priorities will be completely reset in the event of a major breach. Resources currently earmarked for their projects will most likely be redirected to bolster security safeguards if a breach occurs. Make the consequences personal for them. If appeals to historical evidence and reason don't motivate your team, scare them!

- *Manage.* Create a risk register that catalogs and ranks the cyber vulnerabilities faced by the entire IT organization. The register is simply a list of the biggest gaps between an organization's external threats and its internal safeguards. The register should be distributed to the entire IT management team, with individual risks assigned to specific team members. The register should be reviewed and updated on a regular basis by the CIO and CISO. Public awards should be given to staff members who identify new risks and celebrations should be held when risks on the register are retired.

- *Talk.* I had a colleague who used to say: "Anything that interests my boss fascinates me." If IT leaders reference security concerns, activities, or initiatives on a routine basis, these topics will magically start appearing in everyday conversations among team members. Leaders frequently underestimate the extent to which their behavior influences the attention and actions of their peers and subordinates. A simple resolution to reference security issues once a day will have multiplicative consequences throughout the entire IT organization.

- *Measure.* An equally compelling proverb was originally coined by Peter Drucker, the famous management consultant. Drucker said: "What gets measured, gets managed." There are a wide variety of security-related metrics that can be shared among members of an IT management team or across the entire IT organization. Leaders need to identify specific metrics that can engage the attention of the broadest cross section of managers or staff members and then measure, monitor, and report them on a regular basis.

- *Personalize.* Many of the cyber threats that a company faces on a daily basis are similar to the threats that employees face every day as consumers. A deeper understanding of the safeguards employees should adopt in managing their Internet-enabled personal lives is one of the single most effective ways of raising awareness about the safeguards they need to enforce in the workplace. Training programs that help employees understand the steps they should take to secure their personal technology assets will pay major dividends in motivating them to employ similar safeguards at work.

- *Penalize.* In a perfect world, employees would develop an intuitive understanding of the security vulnerabilities associated with their day-to-day activities and embrace the safeguards that have been put in place to protect company assets. Unfortunately, we don't live in a perfect world. Education and management exhortations are necessary but not sufficient to establish a cultural appreciation of security issues. A well publicized framework of graduated penalties needs to be established to ensure that a failure to exercise common sense prudence or respect established safeguards will have consequences for individual employees. Consequences might range from the imposition of additional training requirements, to a preemptive reduction in future merit pay or bonus eligibility, to outright dismissal, depending upon the severity of individual transgressions. Penalties are applied to a variety of other behaviors that threaten workplace safety, product quality, and commercial service reliability. They should be established for actions that threaten information security as well.

Revolutions fail for many reasons. In some cases, the outsized ambitions of their proponents are so disproportionate to the problems they're trying to solve that the revolution collapses under the weight of its own unrealistic goals. In other instances, proponents assume that their goals are so obviously important and beneficial that others will automatically be converted to their cause. This rarely happens in practice. Proponents frequently underestimate the effort required to command the attention of neutral or disinterested colleagues.

The Goldilocks Principle is critical to the success of any cultural revolution. Attempts to over-engineer security crusades with too many policies, too much administrative overhead, too many rules and regulations, and too much management hype will fail. Under-engineered crusades that are planned, scheduled, and budgeted as if they were simply another IT project are also unlikely to succeed. *In the next decade, leaders of successful revolutions will find the appropriate balance between bureaucracy and business-as-usual in designing broadly based security programs that command the attention and enlist the participation of every IT team member. In so doing, they can create a security culture within IT that will serve as an example to their entire company.*

Most Valley firms have a heightened awareness of cybersecurity issues because their commercial operations are hosted in the cloud and they've been forced to deal with the cloud security phobias of their customers. Security safeguards commonly play a prominent role in software engineering activities and the operational management of customer-facing applications and services. Many firms proactively seek external certifications regarding the integrity of their internal security procedures to allay their customers' concerns. The constant addition of new staff members within high-growth companies makes it difficult to maintain the enterprise-wide security awareness that's required to foster a true security culture, but Valley firms typically have a head start on establishing such cultures because their management teams routinely discuss, manage, measure, and enforce security safeguards.

Briefing the Board on Cybersecurity

One of the toughest assignments facing any IT leader is briefing the Board of Directors on cybersecurity. It's a tightrope act. If you're overly confident about the effectiveness of the safeguards you've put in place, you can easily be perceived as being naïve, uninformed, or incompetent. Conversely, if you're too fatalistic about the inevitability of being breached, Board members begin to wonder if they've put the wrong person in charge of their company's security program.

What's the appropriate balance between prudence and paranoia? How can you frame your conversation with the Board in terms they can understand? Here are some suggestions.

Establish a Scorecard

The Board needs some type of framework it can use to keep score on the progress of your cybersecurity program. The NIST Cybersecurity Framework is the most commonly used scorecard but it can easily be altered or extended to address the unique aspects of your company's business model. Progress can be measured in two different ways. You can report on the nature and extent of the safeguards that are being put in place or you can report on the number and nature of the security incidents that are being identified, cataloged, and mitigated.

Obtain External Validation

The Board will want periodic assurances by external experts that your program is properly designed and adequately funded. This is not an indictment of your competence or capabilities. In fact, it's a mark of good judgment to have periodic health checks of your program by acknowledged experts in the field. The Board is going to ask for this anyway, so you might as well beat them to the punch and demonstrate that you welcome external feedback. It's important substantively, symbolically, and probably legally as well.

Report on the News

You should always review corporate breaches that have received media attention since the last Board meeting and be prepared to discuss them head-on. Don't wait for Board members to ask whether similar incidents could occur within your company. Be prepared to discuss any implications of such breaches for your firm and what – if anything – you've done to prevent them.

Become a Professor

Cybersecurity is a complex and rapidly evolving field. Your company's attack surface is constantly expanding through the proliferation of access points and devices. Bad actors are continually exploiting new technologies and vulnerabilities. And you're constantly deploying new tools, skills, and procedures to counter new threats. Every Board meeting is an educational opportunity that should be used to provide the Directors with deeper insight into the challenges you face and the manner in which you're addressing them. The Board is not interested in an academic seminar on cybersecurity but they will appreciate bite-sized updates about emerging threats, new tools, critical skills, best-of-breed operational procedures, evolving regulatory requirements, etc. You've got a tough job – help them understand how tough it really is!

Graduate to Becoming a Trusted Advisor

Once you've gained the Board's respect, go beyond the standard program updates and expose them to information they can't easily obtain elsewhere. Summarize the most provocative presentations or new technologies you encountered at the annual RSA Security Conference. Visit another company of similar size and complexity and report back on the comparative similarities and differences of their security program relative to yours. Share the results of an internal red team exercise and summarize the follow-up actions you've taken.

Boards have matured over time. They're no longer seeking ironclad guarantees that your company is completely protected against all cyber threats. They're simply seeking assurance that you're making prudent investments in safeguards that can effectively address your firm's specific vulnerabilities. The last thing you can do to ease their concerns is to periodically review the Board communication plan that will be employed in the event of a breach or serious incident. The Board may no longer be seeking guaranteed protection of the company's assets but they will want a guarantee that they won't be left in the dark if your safeguards fail.

Compliance

The regulatory environment governing the operations of publicly traded companies is becoming increasingly complex. Financial operations are governed by Sarbanes-Oxley Act (SOX) regulations. Credit card transactions are governed by Payment Card Industry (PCI) standards. Firms operating in Europe must comply with GDPR regulations governing the handling of Personally Identifiable Information (PII). Medical records management must be Health Insurance Portability and Accountability Act (HIPAA) compliant. Firms doing business with the U.S. Federal government may need to be certified by the Federal Risk and Authorization Program (Fed-RAMP). Additional regulations established by the Occupational Safety & Health Administration (OSHA), the Food and Drug Administration (FDA), and the Environmental Protection Agency (EPA) may impact work procedures, supply chain operations, and facility management. In almost every instance, IT systems will play a role in documenting and enforcing the controls required to comply with these regulations.

Compliance is playing an increasingly important role in business-to-business transactions as well. It's becoming increasingly common for large enterprises to demand compliance with PCI or ISO quality standards as a precondition for doing business with their suppliers and partners. Many have developed their own information security standards and routinely require prospective suppliers to complete detailed security checklists prior to procuring goods or services.

The net impact of all of the above is that compliance is no longer a seasonal sport performed on an annual basis to support a company's SOX certification. It has become a year-round activity that must be managed professionally by a dedicated team, with the same level of management attention that is applied to spending practices, project management, and staffing.

Compliance is validated through a series of operational controls. The simplest example of an operational control is the segregation of duties involved in making changes to SOX financial systems. Individuals who develop customized routines to enhance SOX systems cannot be the same individuals who implement such routines in production systems. This control is implemented in practice by denying system administration privileges to the software developers constructing such routines. The control can be audited by reviewing system access logs that identify all the individuals that have accessed SOX production systems during a specific period of time.

Without proper management, control frameworks can grow rapidly. Immature organizations construct unique frameworks for individual regulations or industry standards on a one-off basis. More mature organizations try to maximize the reuse of specific controls to satisfy the requirements of multiple regulations. Effective compliance processes are built around the following control management practices.

Control engineering. Controls are not static. They need to be reviewed and updated on a regular basis. They should be pruned or expanded in response to changes in a company's business operations or its underlying IT infrastructure. Auditors will be particularly concerned about the elimination or simplification of existing controls. The rationale for such changes must be fully justified. Documentation plays a critical role in satisfying auditors that controls have been strictly enforced. Control engineers must ensure that such documentation exists and is readily available.

Control architecture. Compliance controls are frequently an afterthought in major initiatives that introduce new business practices, application systems, or cloud-based services into a company's daily operations. The control framework to be applied to new practices and systems should be considered at the outset of such initiatives, not 4 weeks before their go-live dates! Progressive software engineering teams are increasingly shifting their development practices from DevOps to DevSecOps to ensure that appropriate safeguards are implemented in new systems from the outset, instead of reverse engineering such safeguards into active production environments. This same proactive mentality needs to be employed during the planning of major business or technology initiatives to ensure that an appropriate control architecture is in place when such initiatives go live.

Executive oversight. Control failures can have major business consequences. At a minimum, they may receive Board attention. In the worst case they may adversely impact a company's brand reputation, stock price, or ability to do business with other firms. IT leaders routinely review their spending, staffing, and project plans on a quarterly basis. The coverage, quality, and enforcement of control frameworks should be reviewed on a regular basis as well.

IT leaders commonly treat audits as a pass–fail quiz. They believe that auditors rely on standardized checklists to determine whether an organization is fully compliant with a specific regulation or standard. Although it's true that auditors employ checklists, they also attempt to ascertain the overall health of IT's control environment. Subjective health indicators include such things as the size and seniority of the controls team, the sophistication of control management tools, the level of management attention devoted to controls enforcement, etc. The grade that an IT organization receives at the conclusion of an audit reflects compliance with both the letter and the spirit of a particular regulation or standard. If an organization doesn't appear to be making a concerted effort to comply with the spirit, the formal letter of the regulation will become increasingly important in determining the organization's final grade. It's not uncommon for auditors to deliver punitive findings in their final reports as a means of forcing management teams to embrace the spirit of a particular regulation or standard. Simply put, subjective assurance regarding the importance and integrity of controls management within an IT organization can go a long way towards allaying an auditor's technical concerns regarding the ways in which controls have actually been implemented.

Unfortunately, compliance administration plays to the weaknesses and not the strengths of many IT organizations, especially those within Valley firms. Compliance processes must be thorough (no cutting corners for the sake of expediency); consistent (performed in the same fashion over and over again irrespective of other, more pressing, business demands); and documented (the least favorite task of many IT professionals). *Compliance requirements have simply become too pervasive and business-critical to be shortchanged by oversimplification, attention deficit disorder, or documentation phobias. Dedicated teams will be required during the next decade to ensure that both the letter and spirit of applicable standards and regulations are being consistently enforced throughout every IT organization. Whether IT leaders like to admit it or not, compliance has or will become a critical organizational competency in the next decade.*

Technical Debt

The evils of technical debt are well known to all IT leaders because every organization is afflicted with this disease. It accumulates gradually over time, typically with little notice, and then requires drastic surgery to eradicate.

In the most general sense, technical debt is the gap between the hardware and software systems you have today and the ones you would like to have in the future. Capital expenditures on hardware equipment may be depreciated over several years, making it financially difficult to adopt new generations of technology while the cost of maintaining existing assets increases. Extensive customization of software systems may make future enhancements more difficult and limit the ability to implement new features and functions supplied by software vendors. As hardware and software systems age it becomes increasingly difficult to leverage new products and services that are incompatible with older technologies. In short, technical debt increases operational costs, heightens operational risks, reduces business agility, and retards innovation. It is an evil that distracts IT management, wastes precious staff time, siphons budget dollars from more impactful activities, and undermines the responsiveness of the overall IT organization.

Technical debt manifests itself in many ways. Software systems may contain duplicative or conflicting data definitions, unused data fields, or custom objects performing overlapping functions. Different versions of the same API may be exercised by different objects. Reports generated from different systems may be inconsistent or conflicting. Application response times may lengthen, sometimes notably, provoking user frustration. Regression testing of new enhancements may take longer and require more robust test cases to guard against inadvertent errors. User acceptance testing of new enhancements may need to become more rigorous as well.

Hardware systems may perform reliably over extended periods of time but newer versions of the same technology will inevitably possess smaller form factors, require less electrical power, and be more self-managed than their predecessors. As noted above, adoption of new hardware technologies is difficult to justify when the last wave of hardware investments is still being depreciated.

Effective processes for managing technical debt are somewhat similar to the processes needed for effective weight control. You can either wait for a crisis to develop and then go on a crash diet while joining the local gym, or you can develop good nutrition habits and exercise on a regular basis. In the past, most IT organizations have opted for the crash diet + gym approach to deal with their technical debt. They waited until the cost of debt maintenance became too great or their ability to respond to business needs too restricted before launching a major debt reduction initiative.

Debt reduction initiatives are one of the hardest things to sell to IT's business partners. Such initiatives frequently involve a significant expenditure of funds and a temporary delay in responding to some near-term business needs. Although debt elimination may produce tangible business benefits in the long run, the near-term consequences of debt reduction usually have a limited impact on current business operations. In the short-term, the principal beneficiaries of such upgrades are the IT team itself, which may not constitute a compelling business case in the judgment of many business leaders.

In the new operating model, IT organizations need to adopt good nutrition and regular exercise habits to manage technical debt and avoid its proliferation altogether. The following management practices are essential to minimize and remediate technical debt in the future.

Platform engineering. Software debt tends to accumulate around major finance, human resource management (HRM) and customer relationship management (SRM) platforms, simply because they serve so many different business constituencies. Different functional teams, using their own staff or external contractors, employ different development methodologies to modify major application platforms on a continuous basis. All such changes need to be reviewed and approved by an individual or team that is responsible for the overall health and technical optimization of the platform itself. This is not an architectural function. The platform vendor is responsible for the platform architecture and will likely maintain integrations to other commonly used products and services. This is an engineering activity that seeks to achieve consistency in the ways in which the features and capabilities of the platform are actually being used. Platform engineers should maintain data dictionaries, API catalogs, custom object libraries, naming conventions, and coding standards that can be used consistently by all platform stakeholders. Technical debt has accumulated with little notice in the past precisely because there was no central authority responsible for the health of major platforms. The Platform Engineer role addresses this gap.

Platform engineering concepts are already deeply ingrained within IT infrastructure organizations. They treat their server farms, storage pools and corporate networks as separate hardware platforms that are individually composed of multiple generations of server, storage and network technology. Infrastructure teams typically avoid making wall-to-wall investments in specific versions of server, storage, or network hardware to ensure that they can continually adopt new versions of technology as they become available. They manage each of their hardware platforms as a fleet of assets. They routinely reconfigure existing assets to support different business needs while retaining the flexibility to make seed investments in the latest versions of individual technologies.

Leverage Agile. Software remediation efforts have been launched as grand crusades in the past because there were no viable ways of reducing debt incrementally over time. Agile practices can solve that problem, at least in principle. Software development teams routinely reserve a portion of every scrum cycle to refactor or replace selected components of existing systems. The same development practices can be leveraged by application support teams. Application teams typically work on a series of production issues and enhancements in every scrum cycle. There's no reason why a portion of each cycle can't be reserved to retire some element of technical debt as well. To leverage Agile practices effectively, Platform Engineers should prioritize the debt issues that need to be addressed and business partners need to agree to reserve time in every scrum cycle for debt remediation.

Keep score. Peter Drucker, the internationally recognized management consultant, famously said: "What gets measured, gets managed." Debt problems have been allowed to grow to near unmanageable proportions in the past precisely because IT failed to define and monitor debt metrics. Some of the more prominent manifestations and consequences of technical debt have been referenced above. A more thorough analysis is needed to develop appropriate debt metrics for specific application platforms. It's important to maintain a balance between *manifestation metrics* and *consequence metrics.* The IT staff will be more comfortable monitoring the reduction of manifestation metrics (such as the number of custom objects) but business partners will be far more interested in monitoring the practical consequences of debt reduction (such as page loading times). Debt metrics are desperately needed and should be incorporated in the bonus program objectives of Platform Engineers and application support teams. Ideally, annual performance ratings should be based in part on achieving explicit debt reduction targets.

There's nothing particularly revolutionary about the prescription for debt reduction management outlined above. It's pretty simple: set targets (metrics), make someone responsible (Platform Engineer), and establish a way of avoiding the need for a periodic crash diet (Agile). In the next decade, IT teams that are tired of being constantly victimized by technical debt will institutionalize these three management practices.

Finally, it's important to note that the technical debt associated with individual systems is rarely eradicated in its entirety unless the system is retired and replaced by a wholly new capability. In most instances new debt is continually being created while older, more egregious debt is being eliminated. The true mark of success of any debt reduction process is whether *net debt* (i.e. old debt + new debt – retired debt) is being reduced. Metrics monitoring the creation of new debt are equally important as those quantifying the debt that is being eliminated.

Valley firms are notoriously proficient at manufacturing technical debt within their major application platforms. These platforms address a wide variety of business needs and are continuously modified by multiple functional teams. New fields, objects, packages, and reports may be added to major platforms with little, if any, IT oversight or coordination. Valley IT teams struggle to dedicate resources to the engineering role described above due to other staffing priorities. They also struggle to reserve scrum cycle time for debt remediation due to other, more pressing, business priorities. Most Valley IT teams have application crash diets and gym memberships in their futures!

Workflow Automation

A wide variety of mundane, predictable, repetitive activities occur within every IT organization. Service Desk requests, server patching, data normalization, application testing, cloud capacity management, and many, many other activities can be automated to varying degrees. The benefits of automation are obvious. Automation accelerates the delivery of results, reduces labor costs, and eliminates the risk of human error.

There are a wide variety of automation tools. Some are designed to automate common workflows within specific functional areas such as recruiting, procurement, or order fulfillment. Application vendors offering solutions that support specific business functions are increasingly adding workflow automation capabilities to their offerings or partnering with other vendors who provide such capabilities. This allows customers to derive greater benefits from the use of their applications and also allows the vendors to become more deeply embedded in their customers' daily business operations.

A wide variety of generic tools also exist that are not associated with specific business processes. Generic tools are useful – if not essential – in automating workflows that span multiple business functions which each employ a variety of function-specific applications.

Other tools seek to orchestrate broader, more complex business processes. Orchestration tools typically manipulate automation scripts or software bots that execute individual elements of a broader process. For example, an orchestration engine might assist a consumer in selecting products online based upon their availability, shipping times, and applicable discounts. The orchestration engine might additionally provide the consumer with periodic updates on order status and request feedback regarding the product and their purchasing experience after an order has been delivered.

IT organizations have historically failed to realize the full benefits of workflow automation because generic tools and orchestration engines have not been deployed and managed strategically. Different tools have been procured by different IT teams and applied primarily or exclusively to repetitive tasks within their individual technical areas. It's difficult to develop in-depth expertise in the use of such tools and impossible to achieve material levels of reuse under these circumstances.

Dedicated teams of analysts and engineers armed with standardized tool portfolios are needed if the strategic benefits of automation are to be realized on an IT-wide basis. Dedicated automation teams should be thought of as "headcount factories" that manufacture staff time for higher value work by eliminating time currently devoted to repetitive tasks. Robotic process automation (RPA) vendors have gone so far as to introduce the concept of "digital workers" which they define as the equivalent number of full-time employees that would be required to manually perform the work that is now being conducted through automation. If IT leaders can't obtain the funds required to hire additional staff, automation centers of excellence may be the only realistic way of manufacturing the headcount required to satisfy the ever-expanding demands of their business stakeholders.

This is a particularly propitious time to establish such teams. The majority of generic tools developed in the past employed algorithmic scripts and preprogrammed software bots to perform a series of actions prescribed by human experts. Machine learning technology is revolutionizing the automation tool marketplace and introducing far more advanced capabilities to perform tasks in a less deterministic and more adaptive fashion.

Machine learning doesn't rely upon the prescription of a single expert or team of experts but rather exploits the collective wisdom of groups of individuals who have performed the same task in the past. Machine learning algorithms don't simply generate a particular course of action, they can predict the likelihood that their recommended solution will actually be successful. Risk thresholds can be established to determine when human intervention in a particular activity is warranted and when it is not. Even when human intervention is invoked, machine learning routines can recommend alternative courses of action and stack rank their probability of success, significantly improving the efficiency and productivity of the human agent.

These new tools are far too complex and sophisticated to be haphazardly introduced into an IT organization at the whim of individual teams. They need to be deployed intentionally and strategically. Dedicated analysts and engineers need to be trained in their proper use to realize the full range of potential benefits that they offer.

A variety of organizational models can be used to deploy automation expertise within IT. A standalone center of excellence (COE) can be established within the existing organization to manage the tools, prescribe the templates, and maintain the repository of existing scripts and bots. The COE can serve as a homeroom for engineers and analysts who consult with different IT teams and coach them on the best ways of leveraging existing automation capabilities. Alternatively, automation COEs can be more virtual in nature, consisting of a small team of tool engineers and methodologists with practitioners permanently embedded within individual IT teams.

Regardless of the model you choose, centralized management of tools, templates, and repositories is essential to avoid duplication of effort and accelerate adoption of emerging automation technologies.

To succeed on a sustained basis automation teams need to produce operational benefits that can be measured in terms of time savings, labor reduction, and error elimination. These operational benefits need to be translated into tangible business benefits measured in terms of customer response times, customer satisfaction, staff productivity, cost reduction, etc. Operational benefits should be forecast for each automation project and results should be compared to forecasts. Cumulative time savings should be bookkept, not just on an overall basis but for specific job roles, technical groups, and work processes. In many instances the time savings that an individual automation project can achieve within a specific role, group, or process may be small, but the cumulative savings of multiple projects may be considerable.

Managers play a crucial role in the success of any sustained automation initiative. Managers constantly talk about providing staff members with opportunities to spend more time on higher value work as a means of advancing their careers. When automation projects produce appreciable time savings, managers need to repurpose the responsibilities of impacted staff members immediately. Failure to do so may negate any business benefits that could potentially be realized through the reduction of repetitive work activities. Even worse, failure to expand the responsibilities of impacted individuals may trigger job security concerns. From an employee's perspective, an automation initiative may be nothing more than a thinly veiled exercise in job elimination.

Dedicated automation programs are not trendy, tactical, technical initiatives that ease the workloads of existing staff members. As the IT talent pool shrinks relative to the demand for IT resources, automation initiatives become strategically important. They free up time and dollars that can be used to upskill the entire IT organization. They should be planned and managed in that strategic context.

Workflow automation tools are likely to produce even greater benefits when applied to business processes. IT organizations that establish internal centers of excellence in the use of such tools will inevitably find opportunities to apply their newfound skills to business operations. In the next decade, leading adopters of process automation technology will eventually be able to offer Automation-as-a-Service to their business partners.

Overcoming Automation Indifference

It's been my experience that most IT managers approach automation initiatives with a deep sense of skepticism. Their skepticism stems from the belief that the work performed by their teams is so specialized and so complex that it could never be automated. Alternatively, they may be fearful that they'll be personally reprimanded if business operations are inadvertently disrupted by an automated routine. In the latter case they have a false sense of security that humans perform work more reliably than machines and that they'll experience a loss of control by replacing a human-mediated process with a machine-managed one. Both are false perceptions: machines are actually *more reliable* than humans and managers actually have *more control* over how work is performed when procedures are automated.

As is the case in introducing any new tool or procedure within IT, automation initiatives generally start small. They focus on repetitive workflows of limited scope and marginal business importance. As the benefits of these initial projects are publicized and celebrated by management, automation use cases invariably become more complex.

I created an automation center of excellence (COE) in a prior company that had this exact experience. Initial use cases were focused on improving the efficiency of Service Desk and data center operations. While these projects reduced behind-the-scenes busywork and eliminated certain types of human errors, their benefits were not apparent to our business partners. The results of these initial automation projects – as measured in time savings – were astounding. They far exceeded our intuitive expectations. We actually discounted the results of our time savings analyses because we didn't believe that executives outside IT would believe our results.

News of our early successes spread like wildfire throughout the IT management team. Managers attempted to outdo one another by sponsoring projects that would produce the greatest time savings. Our COE simply didn't have the resources to implement all the ideas being proposed and we had to devise a process for prioritizing automation project requests. Over time we were able to shift our prioritization criteria to place more emphasis on accelerating the speed of our business-facing operations and less emphasis on internal IT time savings. As the initiative progressed, staff members became engaged as well and they would frequently submit project ideas directly to the COE.

Within the span of 6 months, the automation initiative was transformed from a pariah program that everyone supported in principle but no one wanted to join in practice into a popular intramural competition. It succeeded beyond my wildest expectations and was frankly limited only by the number of trained analysts available to work with individual teams. Business leaders started to ask if members of our automation team could assist in automating their internal operations as well.

IT Governance

Perhaps there is no activity within IT organizations that would benefit more from the rigorous application of minimum viable process principles than IT governance. All too often IT oversight committees take on a life of their own and continue to meet long after they've outlived their usefulness. In other instances, oversight committees may spawn a hierarchy of subcommittees or working groups that meet in advance of committee meetings to reach agreement on the information to be presented to their company's business executives. The time and effort devoted to planning committee meetings, preparing documentation, pre-briefing members, and addressing committee concerns can become extremely burdensome. Some business executives would be shocked to discover the total labor hours being devoted to the maintenance of existing governance practices.

Research analysts have pontificated for years about how to "align IT with the business" but experience has shown that there is no one-size-fits-all framework for obtaining guidance and decisions from a company's executive leadership team. An oversight framework that works well in one company may be ineffective or disastrous in another. The formula for successful oversight is complex. It depends upon a company's size, management structure, decision-making culture, executive personalities, and executive preferences. Most importantly, the correct formula at any given point in time depends upon executive perceptions regarding the IT function and its leaders.

It's easy to diagnose ineffective governance processes. Governance practices are failing when oversight committee meetings are deferred multiple times or cancelled altogether. Decisions are made via email. Meeting attendance is delegated, sometimes temporarily, sometimes permanently. Delegation devolves predictably over time following the standard RACI role hierarchy: initial committee members are usually Responsible for the success of the overall company; the next generation of members are Accountable for operational results within their respective functions; and finally, meeting attendees are merely being Consulted or Informed about decisions that are being made elsewhere. It's remarkable how quickly some oversight committees can degenerate from decision-making bodies into information-sharing forums.

There are five organizational principles that need to be considered in constructing an effective governance framework.

Purpose. This may seem obvious but why is there an IT governing body in the first place and what is it seeking to accomplish? Is there a concern that IT is spending too much money or pursuing technology initiatives that have marginal business value? Is IT being asked to do too much and does it need help in prioritizing business demands? Is the current decision-making process for IT investments too ambiguous and ad hoc? Are major initiatives being approved on a one-off basis by the company's CEO and CFO, infuriating other executives who are trying to understand why their IT initiatives are not being funded? Is the company suffering from consensus paralysis requiring inordinate amounts of time to reach agreement on the initiation of major IT initiatives? Does IT chronically struggle to complete funded initiatives on time and on budget? Is IT failing to deliver routine services such as Service Desk support or application maintenance in a competent and timely fashion? If the answer to these questions is "all of the above," then it's likely that more than one oversight committee or review board will be required to address these concerns.

Scope. The responsibilities of any governing body should address a specific set of management concerns. Governing bodies can function as strategic investment committees or as operational oversight boards. Very few committees succeed at providing both strategic direction and tactical oversight because different levels of business management are generally required to provide meaningful guidance on strategic and tactical issues. Many oversight committees suffer from scope creep. They're initially formed to review business cases for major initiatives and approve or reject specific investment proposals. Over time they may start to receive updates on the progress of initiatives that have been approved in the past. They may even start to receive reports on overall IT spending and quarterly variances relative to budget. IT leaders need to guard against such scope creep because it's likely to trigger the delegation of committee responsibilities described above. The seniormost members of a company's executive team will rapidly lose interest in routine project updates and quarterly budget management. They simply assume – quite rightly – that IT's leaders should be managing such activities.

Participation. Again, participation depends upon the purpose of the governing body. If its primary objective is information sharing, participation might be quite large and involve multiple levels of business management. If it is to function as a decision-making body, smaller is always better. Smaller committees are generally able to achieve a broader airing of opinions and concerns and are also able to reach decisions more rapidly. If the objective is to make decisions, then only the true decision-makers should be formal members of the committee. It's sometimes easier to limit participation if the results of committee deliberations are communicated to a wider audience of interested stakeholders immediately after each committee meeting. Delegation rules need to be clearly defined and enforced. To what degree can participation be delegated if at all? Can a quorum of members make binding decisions or recommendations, or must all members be present to participate in such discussions? Experience has shown that it's difficult to impose rules upon oversight committees after they've been launched. It's easier and ultimately more beneficial to define participation and decision-making rules prior to the launch of any oversight group.

Frequency. The frequency of governing body meetings should be determined by business needs. If the body is designed to make financial investment decisions, then it should meet when financial resources are available. Some companies set aside funds to launch new initiatives any time during the fiscal year. Others establish annual budgets and conduct midyear replanning exercises to reallocate funds in response to changing business conditions. Financial investment committees overseeing the IT function should adopt a meeting cadence that can support these different types of investment opportunities. Oversight committees serving more of an information-sharing purpose might meet more frequently on either a fixed or intermittent basis. Senior executive participation is more likely if meetings avoid scheduling conflicts with major company events such as year-end close, the annual sales kickoff meeting, Board meetings, earnings calls, etc. In either case, meetings should only occur when real work needs to be performed: when either important decisions need to be made or important information needs to be shared.

Sunset clause. The charter of every governing body should contain an explicit sunset clause that will force its members to periodically reconsider its purpose and effectiveness. To be perfectly blunt, governing bodies are usually formed to address trust issues that business executives have with the IT function. If senior executives become convinced that IT leaders are spending money wisely, delivering on their commitments, and addressing the company's most pressing business needs, it's likely that the membership and frequency of governing body meetings will change and the overall governance process can be simplified. A sunset clause forces business executives to validate the utility of the governing body and provides an opportunity to make adjustments to purpose, scope, participation, and frequency that will make the body even more effective in the future.

Governance practices in Valley firms are somewhat simplified due to their small size and rapid growth. Senior executives are more accessible and decision-making processes are more spontaneous and informal. Executives are generally quite comfortable making decisions based on limited information or analysis. Financial conversations are primarily focused on the growth of the IT organization, not on ways in which IT spending can be limited or reduced in the future. Prioritization of major IT initiatives is frequently constrained by the availability of subject matter experts within business departments instead of the availability of IT resources. Business leaders are acutely aware of the operational demands being placed on their teams and are reluctant to launch any initiative that isn't absolutely essential or capable of delivering immediate benefits. Business staffing limitations and change management concerns considerably reduce the time devoted to debating the relative priority of competing IT investments. As might be expected, these advantages erode over time as companies grow, management hierarchies expand, and overall IT spending becomes significant.

Confronting the Process Challenge

Process improvement initiatives are hard. They need to be sustained over significant periods of time to produce meaningful results. They need the support and active participation of a broad cross section of managers and staff members. Finally, the most difficult challenge of all is that they never really end. Improvements in one or more aspects of a specific process will inevitably lead to the identification of additional enhancement opportunities. At best, a series of enhancements or extensions will produce a perceptible increase in the overall maturity of a specific process but there will always be more work to be done.

The foregoing list of critical process competencies is overwhelming. Although most IT shops have existing methods for performing each of these processes, they would readily admit that there's substantial room for improvement in the scope and sophistication of their current practices.

If the list of improvement opportunities is long and the work required to achieve meaningful progress is hard, how should an IT management team go about investing time in building strategic process competencies?

IT teams seeking to re-engineer their internal processes should employ the PAAP principle in structuring their continuous improvement programs. PAAP stands for "Prioritize And Avoid Perfectionism." Areas of improvement can be prioritized in several different ways. They could be ranked in terms of their business significance or the benefits they would deliver to IT's business partners. They could be ranked in terms of the efficiency or productivity gains they could produce within the IT organization itself. They could be ranked in terms of the gap between their current level of sophistication and industry best practices. Finally, they might be prioritized in relation to major business initiatives that IT is currently supporting. For example, companies seeking to expand their international business operations might use that opportunity to upgrade their regulatory compliance capabilities. Alternatively, companies seeking to develop customized user interfaces for their retail applications might leverage that initiative to adopt more formal API management practices to ensure that the customer-facing front ends of such applications can be easily integrated with standard back office applications for billing and order fulfillment.

The second half of the PAAP principle is avoiding perfectionism. This is the bane of many well intended process improvement initiatives. As noted in the preceding discussion, processes do not need to resolve every edge case scenario they encounter to be successful. Their ability to efficiently resolve 60%, 70%, or 80% of the issues or decisions that fall within their scope may still yield huge improvements in business agility and IT staff productivity.

Few IT leaders can argue with the need to re-engineer their existing organizational processes to prepare for the challenges of the next decade. But most are loath to act and commit their teams to the hard work that is required to effect meaningful change. If the sheer logic of making strategic investments in critical organizational competencies isn't sufficiently compelling, leaders should consider the plight of their staff members who are knowingly trapped in a daily purgatory of professional frustration, substituting manual labor or devising one-off solutions to plug the gaps in existing processes. Staff members realize only too well that they are squandering their skills and retarding their personal development by manually bridging the deficiencies in current processes. If improvement initiatives can't be justified purely on business grounds, they should be pursued simply to boost the morale and maintain the dedication of IT's hard-working staff members.

PART III

Technology

"When the winds of change blow, some people build walls and others build windmills."

Chinese Proverb

Most IT leaders spend the majority of their time dealing with tactical issues. Their attention is absorbed by budgeting problems, hiring decisions, project reviews, production support issues, personnel problems, audits, vendor evaluations, contract negotiations, etc. Leaders are expected to manage these near-term concerns and also serve as technology strategists, positioning their organizations to obtain the maximum benefits from innovative new technologies. Unfortunately, the demands of routine business can be so great and the steady stream of tactical issues can be so large that leaders fail to recognize or appreciate the secular trends that are radically changing the ways in which technology will be managed in the future. As the proverb goes, they can't see the forest for the trees!

There is ample evidence of this near-term myopia. Over the past 40 years the IT industry has been radically altered by technology changes, starting with an initial focus on mainframe computing that progressed over time to client–server architectures, internet networking, cloud computing, and mobile devices. Some leaders recognized these transitions earlier than others and exploited them as a source of competitive advantage for their firms. Others waited for such transitions to become well established and eventually became midstream or late-stage adopters, simply to keep pace with the business capabilities of their first-to-market competitors.

The technology transitions listed above don't successively replace one another. They simply broaden the range and increase the sophistication of the technology solutions that can be developed to solve business problems. They also create new management challenges. Those challenges are the focus of the following discussion.

This discussion does not attempt to forecast winning or losing technologies that will emerge during the next 10 years. Rather, it focuses on the secular changes in information technology management practices that are happening today and will accelerate in the future. Every IT leader should be asking themselves how they are preparing to deal with these changes.

Technology – What Is It Good For?

It's so easy for IT professionals to become swept up in manias about new forms of technology that they frequently lose sight of why companies invest in information technology in the first place. Our industry's latest mania is over the phenomenon of digital transformation, although it's quite difficult to get two industry practitioners to agree on what digital transformation truly is. From one perspective it's a collection of intriguing new technologies such as machine learning, blockchain ledgers, IoT, virtual reality, data lakes, microservices, serverless computing, bots, etc. While each of these technologies offer tantalizing capabilities, they are all useless – from a business perspective – unless they can be used to address a business problem or seize a business opportunity.

Many technology revolutions have come and gone but regardless of the *rage du jour* IT groups continue to receive funding from their business partners because they routinely accomplish the following three things. If a new technology can't deliver one of the following benefits, it is of little or no use to a commercial enterprise.

Automate processes. From time immemorial information technology has been used accelerate the execution of tasks and the delivery of information. The first computer, known as the Electronic Numerical Integrator and Computer (ENIAC) was developed by two professors at the University of Pennsylvania in 1946. It weighed 50 tons and employed 18,000 vacuum tubes to calculate artillery firing tables employed in World War II. Every successive generation of information technology since ENIAC has succeeded in whole or in part because of its ability to perform a variety of tasks more quickly and more reliably than human beings.

Democratize data. Most companies are drowning in data. Information technology ensures that the right data is inserted in the right process at the right time to achieve a desired business outcome. For example, agents responding to calls in a customer support center need accurate information concerning the nature, timing, and payment histories of recent purchases to respond effectively to a customer's questions. Data also informs tactical and strategic business decisions. Accurate and consistent data is needed to diagnose current business problems (e.g. decreasing inventory velocity) and evaluate the potential value of future business initiatives (e.g. new pricing strategies). IT can readily implement systems that acquire new forms of data and enrich existing databases, but its real value lies in converting such data into business-relevant information that is readily accessible to decision-makers, staff members, and external stakeholders upon demand.

Reduce user friction. IT can reduce the technology friction experienced by employees, suppliers, and customers in many different ways. It can reduce the number of user commands or queries required to complete a transactional process. It can ask a user to validate existing information instead of re-entering the same information a second time. It can reconfigure a customer interface, making it more intuitive and easier to navigate. It can offer the services of a chatbot to assist in locating the information a user is seeking. There are many, many ways in which user interactions with any form of technology can be simplified, streamlined, or personalized. Reductions in user friction can improve workforce productivity and also boost sales and customer satisfaction.

Any technology-based initiative pursued by an IT organization will likely produce benefits along all three of these dimensions. However, experience has shown that the most impactful initiatives produce transformational results along one key dimension. For example, transforming a 2-week loan origination process into a 10-minute mobile phone transaction will be remembered for the revolutionary difference in process automation that was achieved, even if the mobile application interface was frictionless. Alternatively, a new mobile consumer application that boosts online sales and retail store traffic will be remembered for its ease of use, even if the ordering process has been streamlined and new forms of information regarding product quality and availability are being supplied by the app as well. The former example succeeds largely as a result of process automation while the latter succeeds largely through a reduction in end user friction. IT organizations seeking to burnish their transformational reputations would be well served to pursue initiatives that can produce truly revolutionary advances along one of the three dimensions referenced above.

The Technology Management Operating Model

Application Proliferation

There's no definitive estimate of the total number of SaaS business applications in use at the present time. Chargebee and ProfitWell are two startup companies that provide subscription billing services for SaaS vendors. They performed a study of 6,452 SaaS firms within their collective databases in

2019. Crozdesk – a European exchange that connects SaaS sellers with SaaS buyers – maintains a listing of over 19,000 firms within its vendor database. The true number of SaaS business services may lie somewhere in the middle of this range but the more striking conclusion is that there's a huge number of SaaS tools that are potentially available to any modern enterprise.

As the variety of choices has expanded, the barriers to SaaS acquisition have declined. Functional business teams commonly procure SaaS tools without any IT involvement. It's equally common for business teams to select a SaaS vendor on their own and then instruct IT to procure the vendor's product using contract terms and pricing that they've pre-negotiated. In many instances, there's no formal business case justifying the acquisition of such tools and no formal change management plan in place to ensure that the new tool actually achieves its intended results.

During the next 10 years, most companies are likely to employ 5–10 large SaaS platforms to address their core business needs regarding the management of finances, human resources, sales and marketing, supply chains, manufacturing, distribution, retail store operations, customer support, etc. This suite of core platforms will be augmented by hundreds of narrowly focused niche applications that provide complementary capabilities. IT is likely to play a primary role in maintaining core platforms because of their business criticality, technical complexity and cross functional interdependencies. However, the niche applications that make up the long tail of this SaaS-dominated portfolio are likely to be procured and administered by functional teams with very little IT oversight.

The long tail of niche applications creates a management dilemma for IT. IT is ultimately responsible for ensuring that the appropriate safeguards are in place to protect sensitive business information irrespective of where that data resides. IT is also responsible for ensuring that the data employed in daily business operations is accurate, consistent, and timely even if it doesn't have direct control of all data sources. Finally, in a more strategic sense, IT is responsible for ensuring that the enterprise is obtaining the maximum business value from its collective investment in the entire application portfolio *including those components that it didn't purchase and doesn't maintain.*

The principal control mechanism that IT employed in the past to manage a company's application portfolio was cost. IT commonly funded the implementation expenses and subscription fees required to put new systems into production. In other instances, functional teams were required to fund initial implementation costs and IT assumed responsibility for recurring subscription and maintenance expenses in subsequent fiscal years. In either case, IT's financial responsibilities gave it an oversight role and degree of control even if it wasn't directly responsible for the maintenance and administration of every application employed within the corporation.

In a SaaS-dominated world, this cost control mechanism no longer exists. It's far easier – and some might argue preferable – to allow functional teams to buy and maintain niche applications that are uniquely suited to their needs. Although such applications can create potential security and data management liabilities, they are rarely business-critical. It's unlikely that business operations would be materially impacted by the failure of narrowly focused niche applications for a period of several hours or even several days. The outage of a niche application might inconvenience users or embarrass the corporation but the impact of such an event on corporate revenues or profits is likely to be small or negligible.

How can IT fulfill its security, data management, and business value responsibilities when it neither pays for nor manages a significant fraction of a company's overall application portfolio? What type of governance and oversight principles need to be established if IT no longer owns or controls every application supporting daily business operations?

Differentiate systems of record from systems of engagement

Systems of record are applications that contain critical information required to perform essential business functions such as financial reporting or supply chain replenishment. They're also used to demonstrate compliance with externally imposed regulations or standards such as SOX, GAAP, GDPR, or PCI. Systems of engagement are applications that enhance employee productivity or enable interactions with external suppliers, partners, or paying customers. Some applications may serve as systems of record and engagement. For example, a system that consumers use to download movies may recommend future purchases on the basis of a customer's recent buying behavior (engagement) and also log usage credits within a company's loyalty program that can be used to discount future purchases (record).

Enterprises try to avoid the creation of multiple systems of record. Duplicative or inconsistent business information inevitably creates confusion and usually spawns a series of manual processes to resolve discrepancies. Systems of record should be selected through a formal evaluation and approval process involving both business and IT stakeholders. Business requirements should be explicitly defined; vendor capabilities should be rigorously evaluated; and trial implementations should be performed before a final decision is made. Introducing a new system of record is usually a substantial "roll of the dice" for any corporation. The decision is too disruptive, too costly, and too impactful to be made without formal IT involvement.

IT needs to retain responsibility for the maintenance and administration of systems of record. In some instances, these responsibilities may be required to satisfy regulatory concerns regarding the segregation of duties between system users and system administrators. In other instances, IT is the preferred or only group that has the skills needed to manage the technical complexity or functional dependencies of such systems.

IT's involvement in the selection and maintenance of systems of engagement should be more selective. Many engagement applications are designed to improve the productivity of individual employees or work groups. Multiple applications currently exist for workflow management, file sharing, document co-authoring, project management, recruiting, meeting planning, texting, and even email. As labor costs become an increasingly larger portion of a company's overall expense budget and suitable talent becomes harder to find, investments in duplicative engagement tools that improve the productivity of individual teams may be highly beneficial and even desirable. IT involvement in the selection and management of such tools may add very little business value and be difficult to achieve as well. On the other hand, IT may need to assume a leadership role in managing engagement tools that facilitate interactions with external stakeholders such as suppliers or customers because such tools may require extensive integration with other systems or pose potential security liabilities.

Follow the money

IT can't provide any form of governance over applications it knows nothing about. IT groups need to maintain accurate, up-to-date inventories of all SaaS tools currently being employed to support business operations, irrespective of whether they're paying the bill or providing support for every application. This may be a daunting challenge even in relatively small companies. Some business teams simply don't bother to maintain current listings of the applications they are currently using. Others maintain incomplete lists and readily admit they're not aware of all the applications being used by their colleagues.

One of the primary motivations for establishing a comprehensive application inventory is cost management. At a minimum the inventory should contain information regarding the annual cost, renewal date, and business owner of individual applications. Application owners are expected to play key roles in leading or supporting renewal negotiations, ensuring that negotiated price increases and future utilization forecasts are reasonable. If functional teams initially resist application survey initiatives, IT should enlist the aid of their company's CFO. CFOs are always interested in discovering the total amount of software spending within their companies, as well as the spending levels of individual departments.

Application inventories can be constructed in a variety of ways. Substantial information can be gathered through direct interviews with the operations teams embedded within specific functions. Cloud Access Security Broker (CASB) tools may also be useful. CASB tools provide detailed information concerning the websites visited by company employees and may assist in discovering applications employed by relatively small teams.

In summary, IT needs to serve as an information clearinghouse regarding the cost and composition of their company's application portfolio, even if it has no financial or technical responsibility for some or most of the tools within the portfolio. Application inventories need to be maintained on an ongoing basis to ensure their completeness and accuracy. Application owners should be required to validate the information for which they are responsible every 3 to 6 months.

Safeguard the corporation

Selected SaaS applications will contain critical business information that is subject to external regulatory controls such as SOX, GDPR, PCI, or HIPAA. The application inventory maintained by IT should also designate the regulatory regimes that govern the maintenance and administration of individual applications, if applicable. Most IT organizations are well versed in the construction and enforcement of compliance controls whereas most functional support teams are not. The need to maintain a rigorous and auditable controls environment for applications falling within the scope of specific regulations may provide further impetus for transferring their ownership to IT. At a minimum, the existence of in-scope applications outside of IT should be brought to the explicit attention of a company's Internal Audit and Risk Management teams.

Some applications may contain information that is not subject to regulatory controls but is business-critical nevertheless. A comprehensive application inventory might also note the access controls and authentication procedures that are being applied to applications containing sensitive information concerning intellectual property, merger and acquisition planning, ongoing litigation, sales forecasts, etc. In most instances, applications containing sensitive information should be managed by IT because IT has the tools and expertise required to enforce controls regarding data access and system use.

Provide the glue

SaaS platforms are designed to support a specific set of business processes such as customer relationship management (CRM) or human resource management (HRM). Niche applications that complement major platforms are even more highly specialized. A significant portion of any company's value proposition is its ability to optimize processes that extend across multiple functional domains, such as detecting changes in global consumer buying behaviors that can be used to redirect supply chains and adjust manufacturing plans to ensure that sufficient product is available in all major markets. This type of business agility can only be achieved by integrating business logic and data across multiple applications.

Functional teams benefit from integrations between their core platforms and associated niche applications. However, they tend to derive very little benefit from integrating their tools with applications in other functional departments. Consequently, SaaS integrations that are needed to optimize cross-functional processes are routinely deferred or overlooked altogether.

IT groups must champion the cross-functional integration of SaaS tools if their enterprises are to realize the full value of their current SaaS investments. IT should maintain dedicated teams that can construct and maintain such integrations and not rely on functional teams to perform these tasks. Furthermore, IT should proactively identify opportunities where cross application integration can produce business benefits that far exceed the benefits that can be achieved through efficiency gains in individual departments. If IT doesn't seek opportunities to optimize cross-functional enterprise processes, it's unlikely that anybody else will!

Report on utilization

Business leaders rarely appreciate the full suite of applications being employed within their departments. They have even less intuition about the usage of individual applications. Most modern enterprises employ standard tools to authenticate user access to their internal business systems. These tools are administered by IT. They can provide powerful insight into the way in which individual applications are actually being used.

Routine reporting of the cost, allocation, and utilization of subscription licenses within a functional department may enable business leaders to make better decisions about their continued use and distribution. In many cases IT no longer has the authority to launch application standardization or rationalization initiatives but insights into the cost and usage of individual applications may persuade business leaders to reconsider their utility or future allocation.

The Five Myths of SaaS

Subscription-based SaaS applications represented a drastic departure from conventional ways of buying and maintaining business software when they were initially introduced in the early 2000s. SaaS vendors promised that their tools would require no up-front capital investments; be easy to implement; deliver new functionality on a continuous basis; and require minimal customer support. The SaaS marketplace promised access to best-of-breed business capabilities that would reduce historical dependencies on monolithic ERP platforms. After 15 years of practical experience, several of these promises have proven to be myths.

Myth 1: SaaS offers unlimited freedom of choice that allows you to construct an application portfolio that is uniquely customized for your business.

The freedom of choice afforded by SaaS is certainly true in theory, but doesn't always work out that way in practice. Many companies that have adopted SaaS-first application strategies rely on industry leaders in specific functional domains to satisfy their needs, such as Salesforce for customer relationship management, Netsuite for finance, Workday for human resources, Marketo for marketing, etc. Paradoxically, the widespread adoption of these industry-leading SaaS tools has led to the implicit creation of a new cloud-based ERP stack consisting of semi-standard building blocks that must be integrated by their owners.

Myth 2: SaaS provides maximum exposure to innovation by allowing companies to mix and match best-of-breed applications at will.

This myth is also true in principle, but rarely realized in practice. Once a company has built its sales, finance, and human resource practices on platforms such as Salesforce, Netsuite, and Workday, it's quite difficult to switch to alternative systems, even if an upstart company were to offer demonstrably superior capabilities. IT leaders know better than most how much people hate change. While it's true that the technical barriers to swapping SaaS applications are relatively modest, user resistance to change will remain as formidable as it has been in the past. Let's face it – very few IT leaders possess the political capital and intestinal fortitude required to inflict these kinds of changes on their business partners!

On the other hand, changes to niche applications that address very specific needs are easy to implement precisely because they support narrowly defined processes that are performed by relatively small teams. There are a wide variety of best-of-breed SaaS tools devoted to job applicant tracking, sales lead scoring, travel and expense management, training administration, project management, proposal preparation, financial forecasting, revenue recognition, contract management, etc. Niche tools that address these types of needs can be readily replaced by newer, more innovative solutions. This is the key to achieving true business agility through SaaS: the ability to replace niche applications that complement major functional platforms at will.

Myth 3: SaaS tools are built to "plug and play" and can be implemented in very short periods of time.

SaaS applications can be activated and configured in very short periods of time, typically measured in weeks instead of months or quarters. However, to obtain the maximum business benefits from any individual SaaS tool, it's likely that it will need to share data and provide complementary business logic to other applications that support the same or related business processes. When considered in this broader context, SaaS implementation projects can extend for significant periods of time. Important integrations may still be underway a year or more after a SaaS application has been implemented.

Myth 4: SaaS applications are WYSIWYG solutions that require minimal customization.

Enhancements to internally hosted business applications have historically been laborious and painful. Internal IT teams are usually overwhelmed by requests from their business partners for customized enhancements to internally hosted systems. Business representatives conduct endless debates over the prioritization of their individual requests. Delivery dates are frequently missed because individual enhancements turn out to be much more difficult to implement than originally planned.

SaaS vendors promised to solve this dilemma by delivering new business functionality more frequently, reliably, and painlessly than internal IT teams. However, the *quid pro quo* of the SaaS innovation model is that business leaders can no longer request specific enhancements with the assurance that their requests will be fulfilled. In principle, SaaS customers are forced to live with the enhancements that the *vendor* chooses to implement, which quite obviously are designed to address the broadest needs of their customer base. SaaS vendors provide their customers with "What You See Is What You Get" (WYSIWYG) solutions and hope that the steady stream of enhancements they add to their current products will more than make up for one-off customer requests that they're unable to satisfy.

Contrary to initial claims, companies continue to customize their business applications regardless of whether they are internally hosted or cloud-based. Most of the SaaS platforms referenced above are routinely modified through the addition of custom fields, custom objects, and custom application packages built upon one or more platform APIs. Integrations among multiple SaaS applications or with third party data sources add further complexity. Customization has the same consequences whether it's being performed on a hosted application or a SaaS system. It increases operational complexity and risk; creates the need for more rigorous regression testing as each new customization is implemented; and can adversely impact application performance and responsiveness. These issues have not been eliminated in IT organizations that have embraced a SaaS-first strategy.

Myth 5: SaaS is cheaper.

This is a difficult claim to test. Prior to the advent of SaaS applications, business software expenses were incurred almost exclusively within IT and were fairly well known. This is no longer the case. In many instances software is purchased, configured, and administered by functional teams, making it much more difficult to determine if total software spending and support costs have increased or decreased as SaaS applications have replaced hosted systems.

It's obvious that cloud-native companies operating exclusively on SaaS applications have avoided the expense of building and maintaining proprietary data centers. This represents a huge cost savings relative to older firms that employ a mix of privately hosted and SaaS applications. Unfortunately, organizations that are in the process of migrating to a SaaS operating model rarely realize significant cost savings until they achieve a critical mass of SaaS usage. Significant displacements of hosted systems must occur before companies decommission data centers (in whole or in part) and materially reduce current staffing levels. In the interim, overall spending on business software (licenses + hardware + labor) may actually increase.

IT's Shared Role in SaaS Implementation

Responsibility for evaluating, selecting, and implementing new business applications has always been shared between IT and its business partners. Business representatives supply the subject matter expertise required to evaluate the business utility of prospective systems. IT supplies the technical expertise required to determine if and how a prospective system would actually operate within their company's existing technology environment. In the SaaS-dominated world we find ourselves in today, business representatives increasingly take the lead in selecting new applications and turn to IT at the eleventh hour for assistance in completing the procurement process. IT still has an important role to play in the implementation of new tools, even if it hasn't played a primary role in their selection. Specifically, IT needs to ensure that the following implementation issues are fully addressed *prior* to a final procurement decision.

Adherence to security and compliance policies

Does the new tool create new security liabilities? Do new safeguards need to be put in place to address these liabilities? Does the new tool need to be administered in compliance with specific governmental regulations or industry standards? Does such compliance need to be audited and documented on a routine basis? If so, who will be responsible for implementing and enforcing safeguards and administering compliance controls?

Integrations with other applications

Does the new tool need to exchange data with other applications to achieve its business goals? If so, who is responsible for constructing the necessary integrations? When are they required? Who will support the integrations in the future?

Data quality management and adherence to master data management policies

If legacy data is being loaded into the new tool, who is responsible for devising the quality rules to be applied to that data and who is responsible for physically loading it into the tool? If the data being acquired by the new tool is being reused in other applications, does it violate any enterprise-wide data mastering rules (e.g. does it create a duplicative source of business-critical information such as customer billing addresses)?

Change management

Do business users need to receive training on the capabilities of the new tool? If so, who will arrange for such training and who will respond to "how to" questions from business staff members when they encounter problems using the new tool?

Day 2 operations

Who is responsible for devising and administering the rules that will be employed to ensure the quality of new data being entered into the application? Who will manage fine-grained administrative access privileges within the application itself? Who will respond to business requests for reports from the new application? Will these responsibilities be shared in some fashion with IT or will they become the sole responsibility of the functional team purchasing the application?

License and vendor management

Who will administer the allocation of licenses, monitor utilization, reassign licenses as needed, and manage interactions with the SaaS vendor post-implementation? Who will take the lead in negotiations when the new SaaS service contract comes up for renewal?

Experience has shown that a great deal of confusion, wasted energy, and bruised feelings can be avoided if the responsibilities referenced above are clearly assigned to either IT or functional teams *before* a new SaaS tool is put into production. IT doesn't necessarily need to assume responsibility for all of these issues. It merely needs to ensure that someone has!

Infrastructure Abstraction

Infrastructure management practices have been radically transformed during the past 10 years. Virtualization technologies have created multiple levels of abstraction between the users and developers of software systems and the physical computing resources they employ. Users and developers are becoming increasingly less reliant on conventional IT infrastructure teams to procure and manage physical resources. IT teams that support SaaS applications rely upon their SaaS vendors to manage their infrastructure requirements. Teams that build and support proprietary applications have access to resource management tools that marginalize their dependencies on conventional infrastructure groups or eliminate them altogether.

Abstraction technologies have emerged in all components of the standard hardware stack employed by software users and developers. Although the progressive evolution of these technologies is well known, the collective impact they've had on conventional infrastructure management practices is stunning. This evolution can be summarized as follows:

- *Servers.* Hypervisor technology revolutionized the use of physical servers, enabling individual physical devices to host dozens or hundreds of virtual computing machines. Once IT teams learned how to manage virtual servers within their own data centers, they found it convenient to simply rent virtual servers from public cloud providers, further reducing their dependency upon dedicated physical resources. The most recent step in this trend of progressive virtualization is serverless computing in which cloud providers dynamically allocate computing resources on the basis of user needs instead of requiring users to procure predetermined units of server capacity in advance of their actual utilization.

- *Data.* Data virtualization tools provide integrated access to disparate data stores that are associated with different systems or applications. They provide a single user interface that can be used to read and write data into databases located in different physical locations employing a wide variety of file and record formats. They enable real time access to a wide spectrum of data sources without requiring data replication into standardized formats or a single data warehouse.

- *Storage.* Most major storage vendors have developed virtualization tools that allow their customers to optimize the use of the physical storage resources they have purchased. In some cases, these tools apply to a specific product line. In other instances, vendor-supplied tools may optimize utilization across multiple product lines. Software-defined storage (SDS) tools provide a higher level of abstraction that can be used to manage storage resources supplied by multiple hardware vendors. SDS tools run on commodity server hardware and typically support block, file, and object data formats. They provide more flexibility and can potentially produce cost savings relative to conventional storage area network (SAN) and network attached storage (NAS) management solutions.

- *Network.* Software-defined networking (SDN) provides much more flexibility in designing and managing networks that support multiple systems. SDN substitutes programmable management controls hosted on commodity server hardware for specialized network appliances that possess predefined configuration settings. SDN technology was originally used to improve the agility and performance of data center networks by replacing distributed, device-centric management controls with centralized, software-based controls that were based upon a single, logical view of the overall network. It was subsequently extended to wide area networks in the form of SD-WAN tools that could be used to optimize performance across multiple data centers and physical operating locations. The most recent development in network abstraction is intent-based networking (IBN) which employs policies to automate configuration changes across multiple network segments. IBN policies can be associated with specific business conditions, providing the ability to optimize network performance in response to changing business needs.

The abstraction technologies listed above have significantly reduced the manual effort required to procure, configure, and operate physical computing resources. In many instances, they have eliminated the need for these manual activities altogether. This trend will continue and likely accelerate in the next decade. The next generation of abstraction tools will provide a higher level of abstraction that can operate across multiple components of the hardware stack from a single programmable console. In effect, next-gen tools will abstract the capabilities of the abstraction tools we employ today.

Developers will user next-gen tools to construct self-service recipes for the combination of server, data, storage, and network resources they need to build applications, perform distributed database queries, or run complex forecast models irrespective of whether those resources reside within their company's proprietary data center or the public cloud. It will be easy to modify or extend such recipes by simply adding or deleting a few lines of code. It will also be easy to customize self-service recipes by altering a variety of selectable parameters. Higher order abstraction tools exist today. Their capabilities are sometimes referred to as "infrastructure as code." They will only become more comprehensive and more sophisticated in the future.

Wholesale adoption of DevOps practices will further reduce dependencies on conventional IT infrastructure teams. Operational support responsibilities will reside within DevOps teams, not within the infrastructure organization. Infrastructure experts may provide consulting advice but DevOps team members will have primary, if not sole, responsibility for diagnosing application performance issues and making unilateral changes to the virtual computing resources they employ.

Infrastructure abstraction has profound implications for the cost and skill structures of next generation IT organizations. Infrastructure spending on data center hardware, software, and labor has conventionally accounted for more than half the total IT budget. SaaS adoption and DevOps practices are reducing these expenses and reallocating portions of this spending to SaaS support and software engineering activities.

In the next decade DevOps teams supporting proprietary applications will need to develop broader and deeper operational skills. They will need to become more proficient in system monitoring, incident and problem management, forensic log analysis, run time performance optimization, capacity scaling, load balancing, etc. Conventional infrastructure teams will need to operate in more of an oversight and advisory capacity since they will no longer have direct command and control responsibilities for many of the computing resources supporting daily business operations. They will need to ensure that security policies are being enforced, data governance practices are being observed, and cloud computing expenses are being prudently managed across the enterprise.

As we enter the 2020s, many SaaS support teams and software engineers have no comprehension of the physical resources supporting their systems. Their understanding of the ways in which these resources operate and interact with one another is rudimentary at best. During the next decade the gap between software knowledge and hardware understanding will continue to widen, ultimately reaching a point at which a "system" is considered by many IT professionals to be nothing more than a compilation of coded routines that mysteriously runs on something, somewhere that is completely unknown.

Cost Control

It's exciting to speculate about the ways in which information technology will be used to disrupt existing business models and transform IT operational practices over the next 10 years. It's far less exciting but perhaps equally important to contemplate the ways in which technology costs need to be managed in the future. Technology spending increases annually in most enterprises as the use of cloud-based services becomes more prevalent and the number of technology buyers multiplies. IT leaders need to develop a new mindset about how to control such costs, not only to avoid waste but to ensure that their organizations are achieving the maximum return on their technology investments.

Legacy cost management principles are trapped in a construction paradigm. IT has historically been adept at building systems and solutions that address the needs of their business partners. Conventional IT projects involve the purchase of software and hardware; the customization of these assets through a series of coding and configuration activities; the development of operational procedures that will ensure the continuous availability of the solution; and extensive testing before new systems are put into production. A cardinal rule involved in constructing any type of technology solution is to carefully estimate the total cost of ownership (TCO), which includes maintenance and support costs that are incurred long after a new system is initially implemented.

Most IT organizations have become proficient in managing the construction and maintenance costs referenced above. Unfortunately, many of the activities that played a critical role in implementing technology solutions in the past are irrelevant in a cloud-first world where application and infrastructure resources can be procured on demand. *Historical proficiency in the planning and management of technology capital expenditures is becoming far less important than the ability to dynamically manage IT expense spending.*

IT leaders need to stop managing construction costs and start managing consumption costs. Many have not developed the organizational competencies required to fully exploit the pricing models employed by cloud vendors. Construction projects of the past were frequently over-engineered in terms of their capacity and operational support practices. This tendency resulted in part from a desire to ensure that newly implemented systems could cope with variable demand and remain resilient during the foreseeable future. It was reinforced by the practical realization that investment funds were frequently easier to obtain during the construction of a new system than during subsequent years of routine operation. Unfortunately, this tendency to overprovision software licenses and infrastructure assets remains in effect in many organizations today. It is being applied to the procurement of cloud-based resources, resulting in unnecessary expenditures.

Subscription models for SaaS applications have become increasingly flexible and granular during the past 10 years. Various types of licenses exist, depending on the pricing practices of individual vendors. Specific examples include:

- So-called "named licenses" assigned to specific employees or contractors by name.
- User category licenses which may be priced differently for individuals who may have read-only access to a SaaS service versus others who can enter and modify data as well. User categories may also reflect job roles. Licenses for retail store workers or manufacturing plant workers may be priced differently than those allocated to knowledge workers at corporate headquarters. Individual users may or may not be designated by name.
- Daily or monthly average user licenses which are based upon the average number of users employing a SaaS service during a specific period of time. Individual users are not designated by name. Averaging may occur over a period of multiple days or months to eliminate both positive and negative spikes in demand.
- Transaction licenses in which users are charged every time certain function within a SaaS application is exercised (e.g. an expense report is filed or a hotel reservation is made).
- Resource capacity licenses in which companies are charged on the basis of the computing or storage resources required to support their use of a specific application. Under these circumstances there's no limit on the number of individuals using the application. Costs increase as the aggregate usage of computing or storage resources increases.

Hybrid versions of these models also exist. For example, a vendor might charge a subscription fee for every user of a particular application and an additional transaction fee for performing certain types of activities within the application.

Procurement models for cloud infrastructure services offer pricing flexibility as well. Common examples include:

- Computing resources are typically procured on the basis of server capacity and usage times. Usage times may be billed for periods of hours, minutes, or even seconds. Capacity can be reserved in advance at lower rates or purchased on demand on a spot pricing market.

- Serverless computing models have also emerged in which a cloud provider manages the dynamic allocation of cloud resources and customers are charged only for the resources that their application consumes, rather than pre-purchasing specific units of capacity.

- Storage resource utilization is commonly priced on the basis of data volumes that are either transferred into or extracted from cloud-based data repositories.

There are obviously a wide variety of permutations in the ways that cloud resources can be exploited to support the unique requirements of any given company. Individual companies vary widely in terms of their product offerings, functional requirements, operating locations, workforce composition, etc. Although individual departments or lines of business may have optimized their use of cloud resources from a cost perspective, few if any companies can claim to be proficient in cloud cost optimization on an enterprise-wide basis. Managing cloud resource consumption enterprise-wide will become a critical competency of IT organizations during the next decade. Those that fail to develop this competency will waste precious investment dollars on cloud resources that are unneeded and unused.

Leading-edge IT organizations will not only seek to obtain maximum advantage from existing licensing and procurement models, they will proactively propose pricing models that are best suited to their business needs. For example:

- SaaS buyers may define categories of users based upon frequency of usage. They may be willing to pay more for super-users that access an application several times per day and insist on paying less for users who only access the application one to three times per quarter.

- SaaS buyers may require three or more distinct applications to support a critical business process. Each of these applications may be priced in a different way. Buyers may require the vendors supplying these services to harmonize their pricing practices in a way that is best suited to the needs of the buyer. Perhaps the outcome of the business process itself should be used as the basis of payment for the supporting application services, a practice sometimes referred to as "value-based pricing."

- Companies with global operations may move significant amounts of data around the world on a weekly, daily, or even hourly basis. If the entry or extraction of data is being performed to support a revenue-generating business process, perhaps it should be priced at a higher rate whereas data transfers that are performed purely for purposes of synchronization or redundancy should be priced at a discounted rate.

- Companies whose business models are subject to seasonal variations in customer demand, workforce staffing, plant utilization, etc. may be able to characterize their requirements for selected cloud resources in terms of base demand (roughly constant year-round) and surge demand (short-term in nature) and seek differential pricing for each demand category. Note that some cloud infrastructure vendors such as AWS offer this capability today, but this type of bifurcated pricing model is much less frequently employed by SaaS vendors.

The upshot of the preceding discussion is that IT leaders would be well served to spend as much time thinking about creative ways of purchasing cloud resources as they do about leveraging the capabilities of such services.

It's somewhat embarrassing to admit that in the past most IT groups had limited insight into the utilization of the applications and systems they maintained. Information concerning the identity of application users or their session lengths was rarely recorded or retained in a consistent fashion. Internally hosted applications were frequently deployed across multiple physical servers for purposes of load balancing or redundancy, making it difficult to monitor overall resource utilization at an application or user level. Database utilization by specific applications was only investigated in the event of a service outage or slowdown. All of these factors camouflaged the actual usage of IT resources by individual users and applications.

Cloud computing has revolutionized the insight that can be obtained into the usage of business applications and infrastructure resources. Any attempt to develop sophisticated mechanisms for managing the costs of cloud services will rely directly upon IT's ability to develop accurate and comprehensive ways of monitoring consumption at an individual, departmental, business unit, and enterprise-wide level. This is a foundational competency of every IT team that seeks to exercise some degree of control over its company's cloud expenditures during the next decade.

Progressive IT organizations need to proactively replace their conventional skills in managing the cost of technology acquisition, operations, and maintenance with new skills regarding the monitoring and management of cloud resource consumption. Failure to do so will likely result in a severe case of buyer's remorse as business executives outside IT start to question their company's ability to achieve the savings they were originally promised when they replaced their legacy data centers and internally hosted applications with cloud services. Their remorse may be transformed into an indictment of IT's cost management capabilities if IT organizations don't step up to the challenge of managing consumption at a far more granular level than they have in the past.

The Virtual Workplace

The enterprise workplace has gone through enormous changes during the past 10 years. It will undoubtedly change in profound ways during the next

10 years as well. For many individuals work is no longer performed exclusively on weekdays from 8:00 am to 5:00 pm at a fixed physical location. As technology becomes more ubiquitous and as companies become increasingly dependent on temporary employees, suppliers, franchisees, contract manufacturers, channel partners, and go-to-market alliances, the modern workplace is more of a state of mind than a physical reality. Many jobs can literally be performed anytime or anywhere. Even individuals who work in fixed locations such as manufacturing plants or retail outlets use technology during off-shift hours to review work assignments, check equipment status, coordinate vacation schedules, ensure delivery of critical supplies, etc.

At the beginning of the last decade, IT organizations had near complete control of the devices and networks that employees used to access business systems. Desktop and laptop computers were hardwired into a company's corporate network or docked in stations that had a hardwire connection. Mobile phones were mostly company owned. Bring Your Own Device (BYOD) policies were a novelty at the time. They didn't achieve mainstream acceptance until the mid-to-late 2010s. Mobile device management tools became increasingly popular throughout the decade. They were based on the premise that if IT wasn't able to control an employee's personal smartphone, it could at least control part of it! Virtual private network (VPN) technology was routinely deployed on any remote device that was not directly wired to the corporate network – such as a home computer or personal smartphone – to secure an employee's access to critical business systems.

As we enter the 2020s most IT organizations have been enlightened or coerced into relinquishing many of these infrastructure management practices. Employees seek the same digital rules of engagement in today's virtual workplace as they experience at their local Starbucks or favorite airline lounge. They want the freedom to use any device of their choosing – company-provided, personally owned, or publicly available. They expect WiFi-enabled network access everywhere. And they demand simple, reliable single sign-on connections to the cloud applications they need to perform their jobs, plus access to the social and personal applications they need to manage their non-work lives. That's a pretty revolutionary state of affairs when contrasted with the workplace of the early 2010s.

It's difficult to determine if innovative new technologies such as smartphones, Wi-Fi networks, and videoconferencing have altered workplace behaviors, or if workplace behaviors have themselves created a demand for new workplace technologies. In reality, behavioral changes and technology innovations are inextricably linked. New technologies – frequently consumer technologies – are brought into the workplace where they modify behavior. Modified behaviors, in turn, can create a demand for new forms of technology. The end result is the same: employees will think and learn and behave in very different ways in 2030 than they do in 2020, in the same way that the workplace of 2020 differs dramatically from that of 2010.

Two underlying changes in workplace behavior have accompanied and perhaps stimulated the changes in workplace technology that have occurred during the past 10 years. Work has become more collaborative and it is no longer confined to specific working hours or work locations. Collaboration is a continuous phenomenon that occurs on a 24/7 basis, particularly within global companies. There has also been a continuous trend toward downward delegation of decision-making responsibilities. Individual staff members and first line managers have assumed more responsibility because they're frequently better educated; the technology they employ provides immediate access to business-critical information; and many of the repetitive tasks formerly associated with their roles have been automated away. Consequently, they have the ability, information, and time to assume responsibilities that were typically performed by more senior individuals in the past.

These two trends will continue during the 2020s but are likely to change in the following ways:

- *Expanded external collaboration.* The need and desire for business-related collaboration will continue to expand far beyond the employee-to-employee interactions that commonly occur today. Employees will seek information, advice, decisions, and feedback from third party firms providing products or services to their companies. They may wish to consult members of their personal social networks on work-related issues. They may participate on temporary teams composed of individuals from multiple companies and employ virtual worksites to coordinate plans and share progress. These interactions will involve more than simple requests for information or opinions. They will be substantive exchanges with individuals outside the corporation that ultimately lead to business decisions and commitments.

- *Reduced decision latency.* Operational planning and execution cycles in many businesses are shrinking. Unanticipated reductions in foot traffic within retail stores may trigger discounting and marketing activities during the same day or week that they occur. Unexpected changes in seasonal weather patterns may alter normal inventory stocking practices, particularly for perishable goods. Business agility can only be realized if the information needed to make business decisions is accurate, current, and readily available. As the speed of business accelerates, decision latency may ultimately be determined by the data latency within a corporation. The roles, responsibilities, and behaviors of employees in the next decade will be focused to a large degree on reducing data latency within every company's virtual workplace.

Technology advances are equally likely to reshape the modern workplace during the next 10 years. Technologies that could potentially produce disruptive changes in workplace behavior include:

- *Artificial intelligence (AI) enabled virtual assistants.* Employees may no longer need to submit queries or formal requests to obtain the information they need to perform their jobs. AI-enabled assistants may provide such information proactively. For example, sales managers requesting a ranking of their team members' performance might receive comparable information for similar teams for reference purposes, even if such information wasn't explicitly requested. Inquiries regarding inventory stocking levels in advance of the back-to-school shopping season might be accompanied with information about similar stocking levels during the past two seasons, strictly for purposes of comparison. AI-enabled assistants will attempt to anticipate information that may be of use to employees and furnish it proactively instead of reactively.

- *Machine learning (ML) enabled process automation.* Workflow automation has historically been achieved by deconstructing existing processes into a series of tasks that can be enabled through computerized scripts requiring no human intervention. Many if not most automation routines are based on the work habits of individuals acknowledged to be experts in the conduct of specific activities. ML technology provides a means of crowdsourcing automation routines based upon the behaviors of large numbers of experienced employees. This can improve the sophistication of such routines in relatively short periods of time through exposure to a far wider variety of transactional use cases. ML technology can also be used to monitor the work habits of individual employees and develop automated procedures that are uniquely suited to improving an individual's personal productivity. In the past automation has produced the greatest benefits when applied to highly repetitive processes within specific functional areas. ML will enable the automation of activities that are only performed on a periodic basis or require interactions with multiple participants, such as planning marketing campaigns, launching new products, or preparing for contract negotiations. The knowledge workers of 2020 would likely be shocked if they were to discover the extent to which their daily activities will be automated by 2030.
- *Verbal natural language interfaces.* Apple Siri and Amazon Alexa are personal virtual assistants that respond to verbal questions and commands. This technology is in the very early stages of being used within the workplace, but it's a convenient way of obtaining information without submitting a written query or requesting some type of formal report. Sales leaders are fond of using this technology to check on the status of major deals. Supply chain analysts employ it to perform stock checks on specific Stock Keeping Units (SKUs). Marketing

managers employ it to monitor retail foot traffic in key stores during marketing campaigns. The ease of use of this technology is likely to tap a deep reservoir of latent demand for business-related information that historical reporting procedures have been unable to satisfy.

These behavioral and technology forecasts have significant implications for IT management. As the proverbial walls that surround the workings of a modern enterprise continue to dissolve, IT needs to become progressively more proactive at instituting and enforcing security safeguards. How will data be protected when it is shared with external parties? Can the transfer of such data be reliably detected across the collaboration channels available to staff members? Can the movement of such data after its initial transfer be monitored? Should new methods of mutually assured data destruction be established to ensure that sensitive data does not persist for indeterminate periods of time outside the corporation? Will AI-enabled assistants assume some or all of the access privileges of the employees they support? IT organizations will confront all of these security issues and many more in the coming years.

At the same time, IT needs to expand its data management skills to deliver relevant and accurate data on progressively shorter time scales to consumers that exist both inside and outside the corporation. Data quality and consistency issues that were discounted or ignored in the past may severely curtail the effectiveness of business forecasting models. Data sovereignty and gravity issues may restrict the geographic portability of specific data types, placing additional limitations on the construction of accurate business forecasts. Finally, the sheer volume of data that will need to be managed as IoT sensors proliferate is likely to overwhelm our current data management capabilities. Successful data science and engineering teams will be forced to find ways of resolving and mastering these challenges.

The security and data management challenges of the 2020s workplace will be compounded by continued expansion in the number and variety of devices that employees use to perform their jobs. Work may be performed on treadmill touchscreens, coffee shop kiosks, or autonomous vehicle dashboards. Immersive visualization technologies may provide individuals or teams with new ways of analyzing data or designing products. Virtual reality headsets may enable employees to simulate the outcomes of different business strategies – such as product placement in retail stores during the holiday shopping season – from the convenience of their homes or day care centers. The very term "workplace" is likely to conjure up a very different image in the mind of a 2030 worker than it does today.

Workplace Collaboration – Bring in the Anthropologists!

Internet and smartphone technologies have brought a whole new meaning to the term "social interaction." Everyone who owns a laptop or smartphone interacts routinely (sometimes obsessively) with other individuals they've never met, seen, or touched. Many of us have become quite comfortable sharing information, experiences, opinions, and feelings with people in different time zones, cultures, and occupations via a wide variety of social applications. The social technologies that have had such a profound impact on our personal lives have infected the workplace as well and given rise to a profusion of work-related collaboration tools.

At the beginning of the 2020s employees in every corporation use multiple tools for email communication, texting, videoconferencing, task management, document co-authoring, file sharing, project management, proposal preparation, and a variety of other activities. Individual teams have adopted a specific set of collaboration tools based largely upon the whims and preferences of their most vocal members. There's typically no explicit rationale for selecting one form of file sharing tool over another but once a particular tool has been ingrained into a team's daily work practices it's difficult to substitute an alternative.

A long time ago I contacted one of the leading IT research organizations and inquired about the suite of collaboration tools that would be best suited to support work groups distributed in different locations. I posed a theoretical question regarding the optimum tool suite for a 1,000-person team that was centrally located on a St. Louis office campus versus a comparably sized team that was equally split between Atlanta and Seattle versus a comparably sized team that was equally split between London, Chicago, and Sydney. The research organization was baffled by this question. They had no insight into the ways in which collaboration was actually achieved by groups distributed across multiple work locations, but they offered to provide recommendations concerning the leading vendors in each tool category.

Maybe it's time to stop debating the technical merits of different task management tools or videoconferencing systems and bring in the anthropologists! Anthropology is a science devoted to the study of human behavior, both individual behavior and group behavior. Perhaps IT shops in the next decade should start hiring anthropologists to assist business teams in evaluating and implementing different types of collaboration tools.

Anthropologists would make ideal *Collaboration Systems Analysts* (CSAs). CSAs could provide business teams with formal advice concerning the suite of collaboration tools that is best suited to support their daily work activities. A CSA's recommendations would be based on the ways in which work is actually performed, not on the basis of the technical capabilities of individual tools. CSAs would leverage their formal training in anthropology to optimize the productivity of business teams in much the same way that Business Systems Analysts leverage their knowledge of business applications to optimize the efficiency of business processes.

Ironically, our failure to develop a deeper understanding of on-the-job work practices may actually compromise the benefits that our collaboration tools were designed to achieve. Teams that have strong business interdependencies but few common collaboration tools will likely become frustrated in trying to exchange information, develop plans, track progress, etc. Tools employed by different teams to perform similar types of work frequently have conflicting workflows, inconsistent data definitions, and disparate user interfaces. These conflicts, inconsistencies, and disparities may actually end up undermining the productivity improvements that the tools were intended to deliver.

Automation First Mentality

The benefits of automation have been previously discussed in Part II of this book. Automation accelerates the delivery of business or operational outcomes. It reduces the risk of human error and eliminates the rework resulting from such errors. Finally, it can significantly reduce the manual labor required to execute specific processes.

Automation is not a new phenomenon within IT. In fact, IT has pioneered the use of automation tools to manage many aspects of its internal operations including such activities as software distribution, server patching, software testing, password resetting, incident detection, data normalization, etc. Nevertheless, as business models become more complicated and the technology they employ becomes more complex, the potential benefits that can be derived through automation increase, sometimes exponentially.

Automation procedures have typically been applied to well established processes that can be easily decomposed into a series of sequential actions prescribed by current human practitioners. Most processes are initially created through trial and error. Practitioners develop tribal knowledge about the most efficient means of executing repetitive actions and how to handle exceptional circumstances (a.k.a. "edge cases") when they occur. Practitioners are frequently overwhelmed as the number of process transactions increases, so they train junior practitioners who can handle routine transactions while they focus on the execution of edge cases. In many instances rising transaction volumes ultimately overwhelm the capacity of the junior and senior practitioners, compromising the speed and quality of process outcomes. Under these circumstances some or all steps of a process are typically automated or outsourced.

What if we were to think of automation as a *first resort*, instead of a last resort? What if the period of trial and error was focused on discovering the initial 10%, 20%, or 30% of the transactional use cases that could be immediately automated instead of waiting until the full range of automation opportunities was determined? How much time could be saved if the entire process development effort was focused on the identification of easily automated use cases instead of gaining sufficient experience to differentiate simple use cases from complex ones and developing a suitable nomenclature for classifying everything else in between?

An automation first mentality will be a major paradigm shift for IT organizations that justify the size of their technical teams on the need to retain a wide variety of highly specialized skills to resolve the routine issues associated with daily operations. However, it will also be a survival skill for organizations seeking to expand the breadth of their technical expertise in the face of budget constraints and a shrinking talent pool.

A bias toward proactive automation is not sufficient to institutionalize an automation first mentality nor to achieve its prospective benefits. Dedicated resources – both human and technical – are needed to provide automation services to the rest of the IT organization. A wide variety of tools are available to automate processes of varying scope and complexity. Application vendors frequently embed workflow automation modules within their offerings. Generic tools are also available that are better suited to automate cross-functional processes relying upon the capabilities of multiple IT systems. Dedicated Automation Analysts are needed who can function much like Business System Analysts, selecting and configuring the tools that are best suited to automate specific processes.

Dedicated automation teams should maintain repositories of scripts and bots developed in the past and promote their reuse to the maximum extent possible. In addition, they need to develop credible methods for quantifying automation benefits in terms of time savings, error reduction, or customer satisfaction. These general benefit categories need to be explicitly adapted to the business models of individual companies. For example, error reduction may take precedence over time savings in a pharmaceutical or financial services company whereas customer satisfaction might take precedence over time savings in a luxury consumer goods company. Benefit metrics should not be presented in operational terms that make very little intuitive sense to company executives. They need to be tailored to the business concerns of the executive team.

Structured automation programs may produce a business epiphany among some corporate executives but they're more likely to present a long and difficult emotional journey for many employees whose jobs are directly impacted. Most humans dislike change. It's not unusual for existing teams to resist attempts to automate their daily activities. Executives will be convinced of the benefits of such programs by the metrics referenced above. Employees will be convinced over time as they come to realize the amount of repetitive work they're currently performing and as they obtain firsthand proof of the reliability of newly automated procedures.

Structured automation initiatives bear some resemblance to the early days of cloud computing. Conventional wisdom initially dictated that cloud-based applications and computing resources could never be used to support critical business operations. Defenders of the status quo argued that cloud resources could only be applied to a limited set of business needs. Adoption accelerated over time as the benefits and reliability of cloud-based solutions became more widely appreciated and initial phobias were dispelled. The same is true of automation programs. Initial automation projects are usually quite conservative and tend to be limited in scope. However, as their benefits become more obvious and familiarity with the use specific tools grows, automation project proposals multiply. It's not uncommon for Automation Analysts to be overwhelmed with requests for assistance as such programs mature.

Enlightened IT leaders will embrace automation first initiatives purely on selfish grounds. Very few organizations maintain metrics that capture the amount of rework being performed by their team members. In some instances, rework may be unavoidable but more often than not it's the result of human error, either inadvertent or culpable. Since a significant portion of rework reflects poorly on the performance of both IT managers and team members, it's not surprising that rework is so poorly understood or monitored in most IT shops. Automation initiatives deliver double dividends by eliminating errors that may undermine IT credibility both inside and outside the IT organization and by repurposing labor hours that can be devoted to more productive activities. Leaders should consider launching automation initiatives solely for these selfish reasons.

IT leaders who spearhead comprehensive automation programs will be pleasantly surprised to discover that they've added a significant number of so-called digital workers to their staff over time. Roughly 1 to 2 years after such initiatives gain traction, they may employ dozens of digital workers who would have been difficult, if not impossible, to hire through normal budgeting practices.

Vendor Ecosystem Leverage

There's a curious disconnect between the behaviors of technology sellers and technology buyers at the present time. Sellers are making significant investments in extending the capabilities of their products to achieve deeper integration with complementary capabilities offered by other vendors. Buyers continue to focus almost exclusively on the functional capabilities of individual products in making their buying decisions and are failing to appreciate the value of these investments.

Application vendors are expanding their investments in APIs that link their capabilities to services offered by other vendors operating in adjacent business domains. For example, vendors offering tools that support hiring and recruiting activities commonly establish interfaces to human resource management (HRM) platforms to ensure that job candidate information is automatically uploaded into a company's HR database following the acceptance of a job offer. In some instances, vendors offer more than simple interface connections. They may actually enhance the business logic within their application to produce information that is of secondary importance to them but is extremely useful to one or more of their partners. Salesforce.com pioneered the concept of creating an ecosystem of go-to-market vendor partners. Its success has inspired many other vendors to follow their example.

Cloud infrastructure vendors have taken similar steps to facilitate interoperability among their offerings. All major vendors are able to host containerized applications which provides their customers with the ability to move applications from one cloud to another with relative ease. This allows buyers to seek the most economical solution for their needs and also improves the resiliency of their commercial operations by running duplicative instances of selected applications on two or more public clouds at the same time.

Even though vendors have made significant investments in integration and interoperability, most IT buyers continue to make myopic procurement decisions that are primarily based on the features and functions offered by individual products. Integrations with other products that might be of use in the future are a secondary consideration at best in most current buying decisions. This is a missed opportunity. In the next decade buyers will become much more sophisticated in understanding the scope and depth of a vendor's partner relationships and explicitly incorporate such information into their technology buying decisions.

The proliferation of applications discussed earlier in this book makes a vendor's partner ecosystem doubly important. The existence of such integrations eliminates the need for buyers to establish and maintain such integrations themselves. More importantly, *it frequently enables a buyer to obtain more immediate value from their existing application portfolio at no additional cost.*

Buyers need to devote more due diligence to the nature and depth of the integrations that a prospective vendor has established with its partners, particularly those partners whose products and services are already in use within the buyer's organization. Vendors are prone to making integration claims that may be more superficial than substantive. Such claims need to be validated – either through the experiences of other buyers or through direct hands-on evaluations. Enlightened buyers will make better procurement decisions in the next decade by determining the business value that a vendor can deliver via its ecosystem in both the short and long term.

Annual customer conferences provide a unique opportunity to explore the depth and extent of a vendor's partner ecosystem. Partners that have made the most significant investments in product integration will undoubtedly be present and may feature engineers that can demonstrate the nature of the integrations they've created.

While the functional capabilities of any product offering will always play a primary role in vendor selection decisions, IT leaders are overlooking a significant opportunity if they fail to leverage a vendor's investments in partner integration. In some instances, buyers may even require a prospective vendor to establish specific forms of integration with the offerings of other companies as a precondition for purchasing their product.

Managing Enterprise Architecture in the Next Decade

Most IT organizations in larger companies establish Enterprise Architecture (EA) teams to provide strategic oversight of their company's IT investments. To be more specific, EA teams are generally established to ensure that future investments produce tangible business benefits, avoid duplicating the capabilities of existing IT assets, "play nicely" with existing hardware and software systems, and are cost effective. EA teams have historically served as gatekeepers responsible for the evaluation and selection of new software applications and hardware assets. Small startup companies rarely establish formal EA teams because they have more pressing staffing priorities but as companies grow in size from 3,000 to 5,000 FTEs they typically start investing in dedicated EA resources.

How can EA teams provide strategic oversight of a company's technology portfolio in a world in which business teams are evaluating new

SaaS tools without IT's knowledge and software engineers are assembling cloud computing resources in a matter of minutes simply by manipulating a few lines of code? To survive in the next decade, EA teams will need to shed their gatekeeper roles and focus on ways of harvesting more business value from technology investments being instigated by their business partners. They specifically need to do the following.

Think in Business Terms and Think Big

By definition, functional business teams acquire technology that is best suited to address their parochial needs. Supply chain teams purchase applications that optimize inventory velocity. Manufacturing teams purchase applications that optimize factory utilization. Warehousing teams purchase applications that optimize the use of available floor space and minimize labor costs. You get the picture.

Functional teams rarely consider ways in which their IT capabilities can be linked to those of their upstream and downstream stakeholders. EA teams are uniquely positioned to explore ways of leveraging the capabilities of function-specific systems to optimize broader cross-functional business processes. EA teams should always maintain an enterprise-wide perspective regarding their company's overall operational efficiency, which most functional teams lack.

The classic example of a cross-functional process that's critical to the success of every business is the buying experience of a company's paying customers. Customers interact with companies in many ways. Retail customers respond to advertisements and promotions; peruse product information on a company's website; visit retail stores; make purchases; receive invoices; take receipt of products; and may need assistance regarding product quality, delivery, or billing issues. All of these interactions will predispose a customer positively or negatively towards making future purchases. Each interaction is being mediated by one or more business applications belonging to the product management, marketing, retail store operations, order management, and customer support teams. All too often these functional teams operate semi-independently, subjecting customers to a disjointed and unsatisfactory buying experience.

Close monitoring of customer behaviors may also be used to optimize supply chain, manufacturing, and warehousing processes that occur far upstream from the point of sale. Customer responses to advertisements and promotions; navigation patterns on company websites; in-store traffic patterns; and product purchases, complaints, and returns may all provide feedback on size, color, and pricing preferences that can be used to modify manufacturing and stocking plans. In most companies these types of feedback loops are rudimentary, disconnected, and subjectively filtered by pre-existing biases concerning customer behavior. IT capabilities can be readily exploited to reduce the latency and subjectivity of customer feedback information and obtain much deeper insights into future buying preferences and return-to-market trigger events.

In summary, EA teams can deliver strategic business value by leveraging existing IT capabilities to connect the dots among disparate business processes in ways that accelerate overall business velocity, improve company-wide productivity, or enhance customer satisfaction. Optimization of larger scale enterprise processes can produce business gains that far exceed those that can be achieved by optimizing the internal operations of individual functional departments.

Get More Bang for the Buck

It's not uncommon for business teams to acquire SaaS applications with superior capabilities in one or two areas that address a limited subset of their most critical needs. Many SaaS tools are used exclusively to address specific needs even though they possess a much broader range of business functionality. Repetitive reliance on a best-of-breed buying strategy results in a situation in which many teams are paying for a wide variety of IT capabilities that go completely unused. EA teams can potentially deliver significant business value simply by finding practical uses for these unused capabilities.

EA teams should proactively explore the ways in which other companies are using systems they have in common. They can do this by attending vendor conferences where other customers are present or by simply reaching out to a neighboring company to compare notes regarding their respective uses of one or more common systems. Insight into business use cases that have been implemented in other companies may suggest ways of harvesting greater business value from tools you already own.

Another way of harvesting greater value from existing tools is to identify application-to-application integration opportunities that can reduce manual labor, improve data quality, or accelerate a transactional business process. Functional teams are keenly interested in implementing such integrations if they can improve the efficiency or effectiveness of their internal operations. They're less inclined to build such integrations if the associated benefits are being realized by other functions.

Progressive EA teams will keep the API capabilities of their major vendors under continuous surveillance and identify integration opportunities that can benefit multiple functions. They may even establish integration centers of excellence that can construct such integrations proactively, delivering a steady stream of business benefits to multiple functional departments. CFOs are always delighted to find ways they can realize greater returns on their existing investments. Modest successes in expanding the use of products that are already fully expensed are sure to garner praise from IT's financial partners!

EA teams can also play a leadership role in achieving a greater return on future SaaS investments by examining the ways in which a particular vendor's APIs have been employed by other vendors. For example, presume that Vendors A and B have both developed deep product integrations with Vendor C and that A's and B's products are currently in use within your firm. The purchase of Vendor C's product is likely to deliver more immediate and pervasive business benefits simply because of its deep, pre-existing integration with products you are currently using. EA teams are uniquely positioned to determine the synergies that exist among the partner ecosystems of multiple vendors precisely because they have a cross-functional business perspective that spans their company's entire application portfolio. This $1 + 1 = 3$ buying strategy will become increasingly strategic in the next decade as vendors compete for future business through the expansion of their partner ecosystems.

Collaboration Can Conquer Complexity

Personal productivity tools used by company employees have historically been purchased as suites from major suppliers such as Microsoft or Google. During the past decade there's been an explosion of best-of-breed productivity tools for specific activities such as texting, file sharing, document collaboration, project management, videoconferencing, proposal preparation, etc. Many of these tools can be acquired for free by individuals or at a modest cost by small teams. Employees and teams have exploited this no cost/low cost opportunity to assemble customized suites of productivity tools that they believe are best suited to their needs.

EA teams have historically focused on the acquisition of major business applications or hardware resources. They've paid relatively little attention to the wide variety of collaboration tools being used within the enterprise. As modern companies become more virtual in nature and increase their reliance on suppliers, partners, and customers, EA has a unique opportunity to determine the collection of collaboration tools that's best suited to support the work habits of different teams performing different types of work in different time zones and working cultures. In this day and age, it's foolhardy to suggest that a standard set of tools can be used to support all forms of work-related collaboration, but recommendations regarding the comparative utility of different tools in different situations would be welcomed by many teams that have historically relied upon the personal preferences of one or two team members.

In summary, EA teams have a bright future in the next decade if they can stay focused on enterprise-wide business opportunities. In the past EA teams have been derailed by over-engineering vendor selection and approval processes, debating the technical merits of competitive vendor products and engaging in politically damaging crusades to standardize the use of specific tools and systems within large, complex enterprises. Over-engineered selection processes, self-indulgent technical debates, and standardization crusades are a thing of the past. EA teams of the future will succeed by finding innovative ways of applying technology to cross-functional processes, achieving more business leverage from past and future SaaS investments, and maximizing collaboration among all the teams and individuals that have a stake in their company's future success. In the next decade, most EA teams will be limited by their business knowledge, not by their technical expertise. They need deeper insight into how their business operates and how their employees behave and considerably less insight into the technical inner workings of the products they are purchasing.

Business Process Prescience

IT leaders have historically employed a variety of terms to describe IT's contribution to the success of a commercial enterprise. At its most elemental level IT *supports or enables routine business functions*. IT systems can be used to codify business logic and ensure that business processes are executed consistently and accurately. The collection of applications used to support functions such as marketing or engineering improves the efficiency of individual departments and the productivity of its team members.

IT systems that link business logic and exchange data among systems supporting multiple departments serve to *optimize broader business processes*. For example, order management systems are typically linked to warehousing or distribution systems to ensure that products are delivered to customers within prescribed periods of time. Order management systems may also be linked to a company's manufacturing and supply chain systems to ensure that sufficient inventory stocking levels are maintained to meet future customer demands.

Finally, most IT groups maintain enterprise data warehouses that serve as a source of truth about the current status of key operational metrics such as retail store traffic, inventory stocking levels, or outstanding accounts receivable. Data warehouses maintain historical records of these metrics as well which can be used to detect changes in operational performance or inform future business decisions. In effect, IT employs such warehouses to *report the news on current operations and assist decision-makers in evaluating business strategies*.

During the next decade, IT will be able to do far more than simply automate functions, optimize processes, and report the news. The technologies that will be widely available during the next 10 years will provide the ability to forecast or anticipate future business performance in unprecedented ways. IoT technologies will deliver data on current operational metrics at a scale and level of detail that has been unimaginable in the past. Scalable storage resources will exist in the cloud that can maintain and manipulate petabyte-sized data stores. (Let's face it – we'll be talking about exabyte data stores by the end of the 2020s!) And finally, artificial intelligence tools will provide a revolutionary capability to detect correlations and trends in historical data that can be used to forecast future consumer buying behaviors, supply chain risks, product quality shortcomings, etc.

For example, the color palettes that mothers select for their children's back-to-school clothing in August and September may turn out to be strong indicators of the ski jacket colors that they are likely to select for themselves during the ensuing ski season. Health and safety incidents in environmentally sensitive industries may be correlated with the amount of unused vacation time accrued by individuals in specific technical or managerial roles. Companies operating in these industries may choose to restructure and enforce their vacation policies in different ways to reduce the risk of any future environmental liabilities. Job recruiters may be able to win the war for talent by pairing prospective candidates with existing employees during the interview process based upon the common interests and behaviors that are reflected within their social media profiles and interactions. All of these correlations are purely speculative but they serve to illustrate the new predictive role that IT can play in creating business value in the future.

It's always dangerous to create new buzzwords within the IT industry since most new terms initially generate more confusion than clarity (the term "digital transformation" being the most recent example). But the new technologies referenced above truly create a new and different way for IT to contribute to a company's financial success. Perhaps the best description of this new capability is business prescience. Prescience is defined as the knowledge of things or events before they exist or take place. It comes from the Latin word *praescientia* which is directly translated into English as the term "foreknowledge." The fortuitous convergence of IoT, cloud computing, and AI technologies has created the ability for companies to achieve revolutionary new levels of business prescience that can and will serve as a source of competitive advantage in the future.

Many companies have already achieved significant levels of business prescience. Clothing retailers segment prospective customers on the basis of their age, location, education, and socioeconomic factors. They develop customized marketing campaigns for each segment and closely monitor customer response as measured by social media mentions, website visits, store traffic, and product purchases. Marketing campaigns are dynamically altered or replanned based upon sales forecast models that employ these and other response factors.

Pharmaceutical companies and medical service providers collect and share extensive information regarding the efficacy of their drug and treatment procedures. No amount of clinical testing can account for the wide variety of genetic and environmental factors that influence patient outcomes resulting from these procedures. Even the interactions among multiple drugs and treatment protocols can't be fully tested under lab conditions. Efficacy forecast models based upon the experiences of individual patients can significantly reduce unwanted side effects or wasteful spending on ineffective procedures. As in the case of the retail industry described above, efficacy models are based on highly differentiated patient demographics including such factors as family medical history, past living locations, child birth, smoking and drinking habits, etc. Efficacy models are becoming increasingly personalized as genetic variables are being incorporated in forecast algorithms.

Business prescience will become an increasingly important predictor of business success as markets and supply chains become more global. Predictive capabilities will also become more important as business cycle times decrease. Failure to properly anticipate consumer buying behaviors during the annual Christmas shopping season could prove disastrous. Failure to detect early warning signs of failing retail locations could result in a long-term drain on company profitability. Conversely, early detection of customer churn or stock surpluses might trigger changes in pricing policies or discount plans that help retailers avoid the financial consequences of wholesale customer loss or deeper discounts in the future. If Major League Baseball teams can develop predictive models regarding the effectiveness of individual pitches thrown at specific batters, then surely businesses should be able to become proficient at forecasting product demand for individual customer segments at regional, local, and personal scales!

During the past 10 years application buyers have become increasingly sensitive to the partner ecosystems that prospective vendors have established. Partner ecosystems consist of vendors operating in adjacent or complementary domains that have integrated their product offerings with the services of the vendor that is currently under evaluation. Although many if not most SaaS tools are procured because they offer best-of-breed capabilities in supporting a specific business process (e.g. employee recruitment, travel planning, or expense reimbursement), their longer-term enterprise value is greatly enhanced through their partner ecosystems. If one or more partner offerings prove to be useful in the future, they will be inherently easier to implement as a result of the integration that the partners have achieved in the past.

Partner ecosystems will remain relevant in vendor selection during the next decade. However, a product's predictive capabilities are likely to become equally if not more important. Vendors who have incorporated forecasting capabilities within their platforms are likely to be more useful to buyers in the long run, even if their forecasting algorithms require further development or are not put to immediate use. *Vendors possessing predictive capabilities will be increasingly favored during the next 10 years as the volume of relevant business data and the complexity of business processes increases.*

Broker/Integrate/Orchestrate is the New IT Operating Model

It wasn't all that long ago that leading industry research firms were counseling clients to organize their application and infrastructure teams around the Plan/Build/Run activities associated with major business systems. Analysts argued that each of these activities required a unique set of skills and that it would be most efficient to establish dedicated groups specifically focused on system design, construction, and operation. SaaS applications, cloud infrastructure services and DevOps engineering practices have completely invalidated this organizational model.

Business system teams supporting SaaS applications have few if any infrastructure-related responsibilities. They may monitor application integrity or responsiveness but have no way of resolving performance problems directly. They can only report such issues to their vendors. Business Systems Analysts (BSAs) worked with Solution Architects in the past to turn business requirements into technical specifications that could be consumed by a Build team. In today's world, BSAs are frequently able to configure SaaS applications directly, without recourse to the capabilities of Solution Architects or developers. In a SaaS-dominated world, Plan and Build activities tend to merge and Run responsibilities are minimal at best.

Business system teams supporting proprietary applications have largely adopted DevOps practices in which small working groups assume responsibility for both Build and Run activities. DevOps is based on the principle that developers need to understand the consequences of their design decisions by assuming operational responsibility for the systems they've created. Requirement gathering may be performed by BSAs assigned to such groups or by Product Managers that reside inside or outside the team. Design, coding, testing, and production support responsibilities are shared by group members. Cross-training in multiple skills is encouraged (sometimes demanded). This organizational model is the antithesis of the Plan/Build/Run model that was based upon a strict segregation of design, construction, and operational support responsibilities.

The Plan/Build/Run organizational model has been replaced by the Broker/Integrate/Orchestrate model described below.

Brokering replaces Plan. IT organizations used to gather business requirements and build systems that were customized to support their company's business needs. Increasingly in the 2020s, IT shops are assembling suites of SaaS applications and only building proprietary systems if such systems are needed to satisfy their companies' unique business requirements. IT no longer constructs monolithic systems with known inputs and outputs. Instead it assembles a jigsaw puzzle of SaaS and homegrown systems that can support the work processes required to conduct daily business operations.

SaaS platform vendors have established extended ecosystems by exposing their product interfaces to other vendors offering complementary business capabilities. Enlightened IT shops strive to find ways of optimizing the capabilities of these interlocking ecosystems in ways that maximize business value and minimize expense. Enterprise architecture teams should be constantly exploring ways of exploiting ecosystem synergies among their major platform vendors while minimizing conflicting or duplicative capabilities.

Monolithic ERP platforms of the past were closed systems composed of multiple modules that attempted to provide the widest possible range of business functionality. Innovation was delivered to customers through the extension of existing modules or the addition of new ones. The monolithic platform vendors became the primary, if not the sole, source of innovation for many IT shops. That's no longer the case in a SaaS-centric IT organization.

SaaS platform vendors will continue to innovate their offerings and inform existing customers about new capabilities at their annual user conferences. However, it's become equally important to wander the exhibit halls at such meetings to learn about the capabilities of a vendor's partners. It's also advisable to monitor the investment strategies of leading venture capital firms to remain abreast of the technology trends they are pursuing. New best-of-breed applications are constantly emerging. It's IT's responsibility to broker information about new SaaS capabilities that can replace or extend individual elements of a company's current application portfolio. Paradoxically, IT may even find itself suggesting changes to existing business processes based upon emerging SaaS capabilities – a wholesale departure from past practices in which business partners demanded customized modifications to existing applications that would support their operational procedures.

Integrate replaces Build. Conventional build processes are modeled after manufacturing assembly lines employing sequential steps for requirement gathering, hardware procurement, software licensing, custom coding, data migration, and testing. Cloud resources are procured as preassembled piece parts that need to be configured or interfaced with other tools to deliver business value. Progressive IT shops are replacing assembly line thinking with cellular manufacturing models in which security engineers, data warehousing analysts, and API specialists work in concert to integrate new cloud resources into the fabric of a company's existing technology portfolio.

API management will become more important in the future as IT assumes responsibility for integrating data flows across multiple SaaS applications. API governance practices will become more formal and will likely be documented and enforced by a dedicated, centralized group. Individual systems may be procured by business teams but IT will need to ensure that security policies are being enforced and data governance rules are being honored in establishing linkages among multiple systems.

Orchestrate replaces Run. As IT groups shed their hands-on operational responsibilities, Run activities take on a whole new meaning. IT operations teams have historically spent most of their time trying to tune or fix internally developed systems. In a SaaS-centric world, they're reduced to monitoring end user experience and reporting aberrations to their SaaS vendors. (Ironically, end user experience has always been the primary concern of system users, not the hardware utilization or latency issues that preoccupy the attention of conventional operations teams.) Since they're no longer responsible for fixing operational problems, operations teams need to do a better job of detecting service degradation or forecasting service failures. Historical competencies in the root cause analysis of system problems needs to be replaced with new competencies in application performance monitoring and forecasting from an end user perspective.

Since IT is now responsible for establishing the data plumbing connections among various systems, it will need to explicitly establish rules for defining, mastering, transforming, and synchronizing business-critical data across a company's application portfolio. Such rules should already exist but their breadth, depth, and enforcement will need to expand as SaaS tools proliferate across the enterprise. IT is not simply responsible for establishing application-to-application integrations. It's also responsible for orchestrating data flows across those integrations to maintain the accuracy, consistency, and timeliness of critical information being supplied to business partners.

Finally, IT should consider itself responsible for harvesting the maximum business value from the crazy quilt of systems that have been procured by individual functional teams. Functional departments buy specific applications to optimize their internal operations. They don't necessarily worry about the impact of their choices on other business teams whose processes operate upstream or downstream of their responsibilities. Gains in functional efficiency or effectiveness realized through the acquisition of specific applications may be achieved at the expense of broader processes that span multiple functions. IT is responsible for ensuring that the enterprise is achieving the maximum business value from its collective application investments. It is uniquely qualified to identify areas of overlapping or underlapping application functionality precisely because it doesn't (shouldn't) suffer from the business myopia that afflicts individual functional departments.

Epilogue

This book is intended to be thought-provoking and scary. It's thought-provoking because it highlights secular trends in people, process, and technology management that will reshape the way IT organizations operate in the next decade. It's scary because of the sheer breadth and inescapability of these trends.

Every IT organization will be forced to confront these trends and deal with them in one fashion or another. The dilemma faced by IT leaders is to determine which trends need to be addressed as strategic imperatives and which are simply tactical challenges that can be marginalized or disregarded altogether.

Leaders understand the theoretical importance of making strategic investments in new organizational capabilities. However, they're frequently reluctant to embark on such initiatives because they're able to deliver predictable results by continuing to operate as they have in the past; or because they're too distracted by the crush of daily business and tactical concerns to focus on longer-term opportunities; or because the changes required to adopt new practices appear to be overwhelming or insurmountable. I've personally been hesitant to launch major change initiatives for all three of these reasons, even though I knew they were necessary and would be hugely beneficial.

It's ironic and humbling to observe the changes that newly appointed IT leaders are able to accomplish in relatively short periods of time. It's not uncommon for new leaders to shake up their management teams and bring in new talent. They may demand process documentation or process metrics that no one had ever requested before. They may publicly state their intent to introduce new forms of technology into the organization, technologies that have been discussed in the past but never seriously considered for purchase or implementation. I've experienced this phenomenon personally. My successors at several companies were able to accomplish things that I had advocated but never achieved, such as replacing an aging CRM system, moving production applications to the cloud, or implementing BYOD smartphone policies.

Every leader needs to periodically step back and pretend that they're newly appointed to their current role. They need to perform a SWOT analysis to honestly determine the strengths, weaknesses, opportunities, and threats facing their teams and develop focused action plans that bolster existing strengths, remedy weaknesses, seize opportunities, and eliminate threats. Alternatively, they may choose to perform a Rip Van Winkle exercise in which they imagine the ways in which their teams will need to operate 2 or 3 years in the future and then work backwards to identify the management changes that are needed now to achieve their imaginary end state.

Change is a two-edged sword. It's as much about discontinuing legacy management practices as it is about institutionalizing new ones. As John Maynard Keynes – the famous British economist – once said: "The difficulty lies not so much in developing new ideas as in escaping from old ones." Many (most?) IT shops are mired in old perceptions, old concepts, old ways of thinking and operating. We need to shed many historical notions as we enter the 2020s, such as the notion that our best employees have no other employment options. Or the notion that our system construction skills or data center prowess will continue to be a source of competitive business advantage. Or the notion that a serious security breach "could never happen to us."

Every change campaign needs a clear end state vision that is restated repeatedly by its leaders. The vision needs to paint an explicit picture of the new management practices that will be established as well as a clear statement of the existing practices that will be permanently discontinued. In many instances, organizations may be currently supporting a mixture of modern and antiquated practices. Strategic benefits may be achieved simply by standardizing on more contemporary methods or frameworks and aggressively eradicating everything else.

One of Winston Churchill's lesser known quotes is: "It's perfectly well and good to say that everything has been considered, but has anything actually been done?" This book presents its readers with a smorgasbord of strategic challenges. Each reader needs to determine which of these challenges they can turn into opportunities for their companies, their teams, and themselves. It would be unrealistic to embark on multiple initiatives that address each of the challenges referenced in this book but it would be equally foolhardy (and career limiting) to avoid taking any action whatsoever. Readers need to assess the benefits associated with alternative plans of action and trade them off against the change readiness of their teams and themselves. To paraphrase Churchill, it's time to start doing something!

Abbreviation Glossary

ADM: Application Development and Maintenance
AI: Artificial Intelligence
API: Application Programming Interface
AWS: Amazon Web Services
BSA: Business Systems Analyst
BYOD: Bring Your Own Device (to work)
CASB: Cloud Access Security Broker
CIO: Chief Information Officer
CISO: Chief Information Security Officer
COBIT: Control Objectives for Information and Related Technologies
COE: Center of Excellence
CSA: Collaboration Systems Analyst
CRM: Customer Relationship Management
ENIAC: Electronic Numerical Integrator And Computer
DW: Data Warehouse
EA: Enterprise Architecture
EPA: Environmental Protection Agency (U.S.)
ERP: Enterprise Resource Planning
FedRAMP: Federal Risk and Authorization Management Program (U.S.)
FDA: Food and Drug Administration (U.S.)
FTE: Full-Time Employee
GAAP: Generally Accepted Accounting Principles

GDPR:	General Data Protection Regulation (European Union)
HIPAA:	Health Insurance Portability and Accountability Act (U.S.)
HR:	Human Resources
HRM:	Human Resource Management
IBN:	Intent Based Networking
ITIL:	Information Technology Infrastructure Library
IPO:	Initial Public Offering
IQ:	Intelligence Quotient
IoT:	Internet of Things
ISO:	International Organization for Standardization
MDM:	Mobile Device Management or Master Data Management
ML:	Machine Learning
NAS:	Network Attached Storage
NIST:	National Institute of Standards and Technology (U.S.)
OSHA:	Occupational Safety and Health Administration (U.S.)
PCI:	Payment Card Industry
PII:	Personally Identifiable Information
PIP:	Performance Improvement Plan
PMBOK:	Project Management Body of Knowledge
RACI:	Responsible Accountable Consulted Informed (a role hierarchy)
SaaS:	Software as a Service
SAN:	Storage Area Network
SDN:	Software Defined Networking
SDS:	Software Defined Storage
SKU:	Stock-Keeping Unit
SOX:	Sarbanes-Oxley Act (U.S.)
SVP:	Senior Vice President
TCO:	Total Cost of Ownership
UT-Austin:	University of Texas at Austin
VC:	Venture Capital or Venture Capitalist
VPN:	Virtual Private Network
WYSIWYG:	What You See Is What You Get

Index

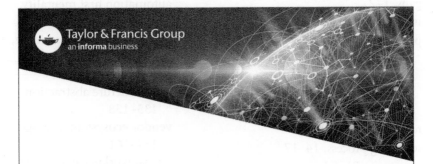